I0127796

Chinese Environmental Ethics

Chinese Environmental Ethics

Religions, Ontologies, and Practices

Edited by
Mayfair Yang

ROWMAN & LITTLEFIELD
Lanham • Boulder • New York • London

Published by Rowman & Littlefield
An imprint of The Rowman & Littlefield Publishing Group, Inc.
4501 Forbes Boulevard, Suite 200, Lanham, Maryland 20706
www.rowman.com

Copyright © 2021 by The Rowman & Littlefield Publishing Group, Inc.

This volume contains the following previously published material, reproduced with permission:

Nicolaisen, Jeffrey. "Protecting Life in Taiwan: Can the Rights of Nature Protect All Sentient Beings?" *Interdisciplinary Studies in Literature and the Environment* 27, no. 3 (September 2020): 613–632. By permission of Oxford University Press.

Sections of chapter 5 first appeared in: Campany, Robert Ford. *The Chinese Dreamscape 300 BCE–800 CE*, Harvard-Yenching Institute Monographs 122. Cambridge, MA, and London: Harvard University Asia Center, 2020.

All rights reserved. No part of this book may be reproduced in any form or by any electronic or mechanical means, including information storage and retrieval systems, without written permission from the publisher, except by a reviewer who may quote passages in a review.

British Library Cataloguing in Publication Information Available

Library of Congress Cataloging-in-Publication Data

Names: Yang, Mayfair Mei-hui, editor.
Title: Chinese environmental ethics : religions, ontologies, and practices / Edited by Mayfair Yang.
Description: Lanham : Rowman & Littlefield, [2021] | Includes bibliographical references and index.
Identifiers: LCCN 2021023700 (print) | LCCN 2021023701 (ebook) | ISBN 9781538156483 (cloth) | ISBN 9781538156490 (paperback) | ISBN 9781538156490 (epub)
Subjects: LCSH: Environmental ethics—China. | Environmentalism—China.
Classification: LCC GE42 .C45 2021 (print) | LCC GE42 (ebook) | DDC 179/.10951—dc23
LC record available at https://lccn.loc.gov/2021023700
LC ebook record available at https://lccn.loc.gov/2021023701

Contents

Acknowledgments

This edited volume started out as an international scholarly conference held at the University of California (UC), Santa Barbara, in January 11–12, 2019. We would like to express our deep appreciation for our funding agencies: the UC Santa Barbara Confucius Institute, the Department of Religious Studies, and the Interdisciplinary Humanities Center. The Confucius Institute also helped with a publishing subvention to allow for Open Access digital publishing, one year after the initial print publication.

We also wish to thank several scholars and graduate students who participated in the conference, as paper presenters, or as discussants: Prof. Chen Huaiyu, Prof. Sonya Lee, Prof. Thomas Mazanec, Prof. Vivian-Lee Nyitray, Sarah Veeck, Julia McClenon, and Susie Wu. Thanks to Dr. He Fang for her valuable staff support in travel and meal arrangements, before she took an academic position at New York University in Shanghai. We also acknowledge the editing and indexing help of UC Santa Barbara PhD students Nate Sims and Yi-yang Cheng.

The editor, Mayfair Yang, also wishes to thank her husband, Philip Myers, for his generous and reliable moral support throughout.

Introduction

Mayfair Yang

乾稱父，坤稱母；予茲藐焉，乃混然中處。故天地之塞，
吾其體；天地之帥，吾其性。民，吾同胞；物，吾與也。

Heaven is my father and Earth is my mother, and even such a small
creature as I, finds an intimate place in their midst. Therefore, that
which extends throughout the universe I regard as my body and that
which directs the universe I consider as my nature. All people are my
brothers and sisters, and all things are my companions.

—Zhang Zai 張載
(De Bary and Bloom 1999: 683)

Glimpses into premodern Chinese thought and culture provide treasure troves
of ancient environmental wisdom and ethics, such as the eloquent epigraph
above by the Neo-Confucian Zhang Zai (1020–1077 C.E.). Here, he melds
Confucian kinship and family ethics with the Daoist emphasis on "primary
breath" (氣 *qi*) as a life force that runs through all living things, the earth, and
the cosmos. Thus, the life force of the universe encompasses and connects all
things within a shared body of kinship relatedness, and human beings are not
elevated above all life forms, but embedded within larger cosmic orders. Tu
Weiming calls this ontological imaginary a Neo-Confucian "anthropocosmic
worldview" (2001). While these environmental gems of thought from the past
certainly did not prevent environmental destruction throughout Chinese history (see Elvin 2006; Marks 2017), they do bring into stark relief the modern
separation of Nature versus Culture, which was not native to China. These
oppositional categories were imported in early twentieth-century China, in
the name of saving and building the nation, revolution, and industrial development. Although Chinese religiosities underwent a traumatic history of

1

suppression throughout the twentieth century,[1] they have bounced back in the new millennium as Chinese people search for ways to deal with the new demands and travails of modern life.

The Chapters collected in this volume explore the possibilities for Chinese religiosities, past and present, to contribute to building new environmentally friendly cultures, values, and spiritualities for our planet. The chapters were first presented at an international conference at the University of California, Santa Barbara in January 2019: *Chinese Religio-Environmental Ethics and Practice*. The contributors form an interdisciplinary group of scholars that includes cultural anthropologists (Adam Chau, Yang Der-Ruey, Liang Yongjia, Mayfair Yang), religious studies scholars (Robert Ford Campany, Wei Dedong, Jeffrey Nicolaisen), a sociologist (Huang Weishan), and a historian (Tristan G. Brown). Some chapters examine primary Chinese texts tracing back to medieval or late imperial China, while others are based on fieldwork, interviews, and research visits in China in recent years. The chapters do not merely search for textual or scriptural sources that are conducive to modern environmentalism but share an emphasis on how these teachings were put into social practice. Some chapters, that is, chapters 4, 6 and 7, also reveal the intractable difficulties of retrieving or remobilizing ancient environmental ethics in the context of modern secular life, where new utilitarian and political realities hold sway, while chapter 8 shows that modern religio-secular nationalism may also give rise to environmentalism.

For a long time, environmental studies were the domain of the STEM fields of natural sciences and engineering, or the legal, managerial, and policy-making professions. However, in almost every country, there seems to be a tremendous gap between our extensive scientific knowledge about the dire nature of climate change, pollution, water shortages, and species extinctions, and any concerted political or social will to change course from our destructive modern capitalist industrial-consumerist ways. The examination of religio-environmental ethics holds the promise that beyond professional elites and their technocratic or legal fixes on environmental damage, we can also explore how to instigate transformations at a deeper fundamental level of the religio-cultural collective unconscious, which will involve a broader participation of people from different walks of life.[2] The Norwegian philosopher Arne Naess (1989) called for a "deep ecology," rather than a "shallow ecology" of techno-scientific knowledge that is walled off from considerations of basic ontology, ethics, cosmology, emotions, rituals, lifestyles, and politics. Promoters of traditional Chinese medicine often say that its effects may be slow, but rather than merely treating the surface symptoms of an illness, Chinese medicine tackles the foundations and roots of a health problem. Since Chinese religious traditions still resonate with contemporary Chinese culture, even among non-religious people, tapping into their environmentalist elements holds the promise of helping to reverse or moderate modern environmental degradations at a deep structural and systemic level. Some of the following chapters embark on a search for environmental elements in religious cultures that still resonate today, which can help us make the changes we need to live *with* the earth, and better coexist with other living things.

Since religions are ritualized, affective, and cosmological elaborations around our human existential realities of life and death, they are a most valuable domain to examine, because our ecological crises are all about the fundamental question of life and death. We as a species now share two new common environmental threats to our lives: global climate change and COVID-19, in addition to the old problems of poverty, warfare, and other diseases. Since human population has expanded to almost eight billion today, with environmental damage and depletion of natural resources, we are encountering the question of death on a massive scale. Death is a deeply religious subject, and since ancient times, it has led to reflective thought and the evolution of social rituals and institutions to explain and deal with the pain, fear, and suffering it causes. Religiosity is a rich field to mine for alternative modes of being that may help us to extract ourselves from our current self-destructive forms of subjectivity and social motivations. Given the grave, ongoing drastic extinction of other species, our dominant ontological structures of being human in relation with other life forms and the earth need to be readjusted.

I. SIGNS OF THE ANTHROPOCENE IN THE UNITED STATES AND CHINA

Our 2019 conference coincided with the fiftieth anniversary of the historic oil spill off the coast of Santa Barbara, California, when a blow-out occurred on January 28, 1969, on Union Oil's oil drilling Platform A in the Dos Cuadras Offshore Oil Field. Within a ten-day period, this became the largest oil spill in the United States at that time,[3] with an estimated 80,000 to 100,000 barrels (13,000 to 16,000 cubic meters) of crude oil spilled into the Santa Barbara Channel and onto the beaches from Goleta, through Santa Barbara, to Ventura. Countless marine wildlife were coated with black oil and tar, including elephant seals and sea lions at a World Heritage Site on the western tip of San Miguel Island, and unknown numbers of fish, dolphins, and sea birds perished. The public outcry and media coverage were so extensive that the oil spill contributed to the establishment of the Environmental Protection Agency (EPA) in 1970 in Washington D.C., and to Congress passing the Clean Water Act in 1972, the Endangered Species Act in 1973, and the California Coastal Commission in 1972 to protect California's coastline. On the part of civil society, local citizens established the Environmental Defense

Center, and organized the world's first Earth Day in Santa Barbara on April 22, 1970, now with the global participation of 190 countries. UC Santa Barbara, my university, opened the country's first undergraduate Environmental Studies Department in 1970.

During the Trump administration, many of those early environmental policy gains in the United States were seriously eroded and threatened by industrial corporate lobbyists. The Trump administration transformed the EPA into the *defender* of corporate industrial and fossil fuel interests. It dragged its feet in properly dealing with the two most perilous environmental threats to the planet: climate change and the COVID-19 pandemic. It weakened or repealed at least one hundred U.S. environmental laws and policies,[4] suspended government regulations on the release of methane and toxic waste from coal plants into public water sources, and mounted efforts for oil companies to drill in the Arctic Wildlife Refuge of northern Alaska's wilderness, on Federal protected public lands, and in Federal waters three miles out from the California coast. It has been heartening to see the Biden administration's robust environmentalist agenda and the decision to rejoin the Paris Climate Accords. In the past decade, California and other Western states have also experienced increasing droughts and wildfires, whose new intensities have alarmed firefighters, brought devastation to homeowners, and startled insurance and electric power companies. In the summer of 2020, there were over 8,200 fires that burned four million acres across the state, and much of the state was blanketed in lung-damaging smoke. Seven out of the ten most destructive fires in Californian history have burned in the past five years, and the total land burned in the August Complex Fire of 2020 was larger than all of the recorded fires in California between 1932 and 1999.[5]

During the Maoist era (1949–1976) in China, a top-down state command economy fashioned after the Soviet model, mobilized the national will to industrialize and conquer nature. The centralized state's exhortation, "Humans must be victorious over Nature!" (*ren ding shengtian* 人定胜天!)[6], gave expression to human revolutionary will overcoming nature, no matter the local terrain, climate, or vegetation. In her book, *Mao's War Against Nature*, Judith Shapiro (2001) describes state mass mobilization campaigns to assert human dominance over nature, including efforts to increase human population and the reclamation of wetlands and lakes for farming or towns, which damaged fragile ecosystems. The conversion of Mongolian grasslands into wheat fields exacerbated northern China's desertification. The "Wipe Out the Four Pests" campaign killed sparrows in great numbers, only to see an increase in the insects that the birds ate. True to the imported Western linear history and social evolutionism, the nation threw itself into making a "Great Leap Forward" (1958–1961) in economic advancement. This mass mobilization campaign led to deforestation for firewood to feed the backyard furnaces that produced low-grade iron in the steel production drive. Thus, by the time of my first visit in 1982 to sacred Mount Tai in Shandong Province, home of the largest temple of the Goddess named Sovereign of the Azure Clouds (碧霞元君 Bixia Yuanjun), I was shocked at the towering pile of almost bare dirt and rocks. This landscape denuded of trees was an example of what Anna Tsing (2017: G2) and others have called a landscape "haunted by past ways of life," deprived of its gods and goddesses, its ancient forests, wildlife, and insects, who have all become "ghosts."

The state command economy seems to have been just as environmentally destructive as liberal capitalism: while liberal capitalism is a system for private companies to plunder the environment for profit, state economies were motivated by expanding and glorifying state power. Whatever the motivations or tactics, they share a common Enlightenment faith in a linear history of "progress" or "revolution," economic and technological developmentalism, and a discourse of human "liberation." In the post-Mao era, with China entering global capitalist systems, state and capitalism are fused, producing unprecedented rapid economic development. In hindsight, the Maoist insistence on materialism, evolutionism, and individual self-sacrifice for the nation or the revolution, and the rejection of older spiritualities, religious ethics, and rituals, together laid down the bittersweet foundations for China's economic success, consumerism, and severe environmental degradation today.

In the four decades of post-Mao economic reform, China has rapidly built up a relentless economic engine of industrial growth, export economy, urban expansion, and a powerful science and technology regime. While hundreds of millions of Chinese people were lifted out of poverty, a very impressive record, they also endure extensive pollution of air, water, and soil, and thousands of cancer hotspots across the country (Shapiro 2016; Smith 2020). Perhaps no other nation has experienced this rapid industrial and economic development and infrastructure-building on such a massive scale. China is no longer considered part of the "Third World," for its middle and upper classes enjoy considerable wealth. These developments are all the more amazing when one considers that in the nineteenth and early twentieth centuries, China, like India, was regarded by the West as a place hopelessly mired in tradition and religion.

China has now joined the West as the planet's biggest perpetrators of climate change, pollution, and environmental destruction. In 2017, China's total emission of CO_2 from the vast amount of fossil fuels it burned was Number One in the world, at 30 percent of global emissions, while the United States ranked Number Two. In 2012, China's production of methane gas, which is even more harmful to the climate than carbon dioxide, was 3.5 times that of the United States. Especially polluting is China's continued reliance upon coal, most of it unprocessed or uncleaned "dirty coal," which in 2017 still

generated 70 percent of China's electricity. In 2020, even more coal-fired
plants and electricity generators were planned or built (Frohlich et al. 2019;
Smith 2020; Chai 2015); however, coal-fired generators now produced 60
percent of China's energy.[7] Chinese fishing fleets now represent the largest
number on the planet, propelled by state subsidies, and fishing in far-flung
and unregulated international waters. They are often seen pulling out the rich
marine life around the protected waters of the Galapagos Islands and depleting the seas around Antarctica.[8]

However, we must also remember that the Western nations of Europe
and North America long contributed to climate change since the Industrial
Revolution, and if we consider the *per capita* consumption of fossil fuels,
consumer goods, and meat today, the United States is at the top. Furthermore, China's alarming contribution to climate change does not just represent
consumption by its own people, for Chinese manufactured goods are mainly
exported to the world. Thus, the whole world is responsible for China's
greenhouse gas emissions, especially the global rich and middle classes. The
juggernaut of Chinese industrialization and its global impact means that all
the more, we need to figure out what elements in Chinese religious traditions
we could develop to counter these modern environmental threats.

In 2020, Wuhan City residents underwent the traumatic experience of a
severe 2.5 months lockdown of the entire city of nine million people to prevent the spread of the deadly COVID-19 virus. In February, when I helped to
organize a shipment of nine pallets of N95 masks and medical protective suits
by our Santa Barbara global disaster relief nongovernmental organizations
(NGO), Direct Relief International, to the city of Wenzhou, a COVID-19
hotspot, we had no idea that our own country would soon have the greatest
number of COVID-affected cases and fatalities in the world. As scientific
evidence suggests, COVID-19 is likely the result of zoonotic transmission
from bats to humans, most likely through an intervening species.[9] It joins
a number of other zoonotic diseases that have jumped from nonhuman species to humans in modern times: HIV, Ebola, SARS, MERS, avian bird flu,
Zika, and Mad Cow Disease. The reason for the increased frequency of these
new epidemics is the growth of human populations, their encroachments
into wildlife habitats, and increasing commodification and consumption of
wildlife, whether through wild capture or wildlife farming. Zoonotic diseases
that leap across the species barrier can easily become contagious and spread
in wet markets, where wildlife species are crowded together in tight cages
and under stress, infecting the humans who slaughter, clean, and cut them up.
Thus, the coronavirus pandemic is very much part of our growing environmental crisis of humans causing the loss of wildlife habitat.

The summer of 2020 was also the time of great flooding in China, with
unprecedented torrential rains that came down across central and southern

China, raising water levels to historical highs in 433 rivers.[10] Several dams and levees were blasted apart by local governments to redirect the rising waters away from heavily populated areas. The floods affected and displaced over sixty-three million people, and many lost their homes and belongings. There was also the danger of huge crop losses leading to food shortages and industrial production shutdowns throughout the areas of major rivers in central China. Not since 1981 have the rains been so severe across China, while Bangladesh and eastern India were also flooded. Intense anxiety focused on the stability of the Three Gorges Dam on the mighty Yangzi River, the greatest hydraulic engineering project in the world, and a point of Chinese national pride. People also worried about the welfare of the giant cities of Chongqing upriver, and Wuhan and even Shanghai downriver. Satellite images of the dam before and after the first three flooding events of summer 2020 led to some global speculation that the dam had suffered some buckling due to the surging waters. Indeed, Xinhua News Agency on July 18 admitted that the dam had experienced "displacement, seepage, and deformation."[11] Yet, even as the Three Gorges Dam has withstood the flooding, there are 94,000 smaller aging dams in China, most of which were built in the 1950s to 1970s and not always maintained well. During the Maoist era, revolutionary zeal and mass mobilization greatly altered the landscape of China with reclamations of marshes and lakes and state infrastructural engineering feats such as dams, highways, and bridges.

Given these grave environmental events in 2020 in just two countries, the growing global discourse that our planetary life has entered into the geological epoch of the "Anthropocene" helps to account for these novel extreme environmental events. First coined by the atmospheric chemist Paul Crutzen (2002), the Anthropocene is, according to Dipesh Chakrabarty (2009), the epoch when human beings are no longer merely biological organisms or just social animals, but are now a "geophysical force." Our species' social activities have now so seriously altered the geography and biology of the planet, and changed the chemistry of the earth's atmosphere, that "natural forces" are no longer as independent or separate from human forces. Although some trace the Anthropocene to the Industrial Revolution, and others to the beginning of agriculture ten thousand years ago, in 2019, the Anthropocene Working Group[12] voted to set 1945 as the start date for the Anthropocene (Subrmanian 2019). In that year, the first atomic bomb blasts lodged radioactive debris deep into sedimentary layers and glacial ice around the globe, providing geological markers that signaled the crossing into a new threshold of human-induced climate change. The postwar 1950s was a period of huge population growth, when more societies across the globe industrialized, expanded agriculture with heavy chemical fertilizers and pesticides, increased transportation relying on fossil fuels, saw the industrial production of meat and poultry, and some nations experimented with thermonuclear weapons explosions.

As shown by the effectiveness of China's response to the COVID-19 pandemic, the modern Chinese state can also serve as a key leader and organizer of environmental protection. During the Maoist era, the state deployed the religio-revolutionary fervor and nationalism in the state mobilization of tree-planting across the nation by the masses. Adam Chau's chapter 8 shows that Maoist era state–sponsored tree-planting drives, whether at the national, local, or work unit levels, are still carried on today, and have become "intermeshed" with temple-based folk environmentalism and transnational environmental NGOs. No other country in the world experienced the systematic human population control that was instituted from 1979 to 2015. Although the draconian population policy often relied on harsh methods, bringing suffering to families, mothers, and children, from what Dipesh Chakrabarty called a "planetary" standpoint (2019), it also tempered human population growth, which is fueling the alarming mass extinction of nonhuman species. This population control was so effective, that today, policy makers in China worry about their country rapidly following in the footsteps of Japan and South Korea as aging societies of low birth rates. In recent years, China has also become the center of the world's green technology manufacturing. In 2017, China's global share of solar cell production was 68 percent, while solar panels were 70 percent of global manufacturing (Finamore 2018: 68–70). It has also installed huge solar and wind power capacities across the country, and subsidizes the rapid development of "new electric vehicles" (NEV), such as passenger cars, electric buses, and motorcycles. Its economy of scale has been a gift to the world, as China has brought down the costs of green technology around the world: a drop of 72 percent for solar power, and 67 percent for wind power (Finamore 2018: 72).

II. ENVIRONMENTAL ETHICS: BEYOND
THE ANTHROPOLOGY OF ETHICS

This volume seeks to address a *lacunae* in both the fields of environmental humanities, as well as the anthropology of ethics. What is missing in the former field is a recognition of the importance of *religiously* inspired environmental ethics or ontology, while the latter field tends to narrowly focus on *human* ethical practice. While the anthropology of ethics follows modern Western philosophy in focusing primarily upon the ethics of human relations, our volume seeks to broaden ethical inquiry to environmental and interspecies ethics. This volume aligns with James Faubion's distinction between "ethics," which lead to change and restructuration, and what he calls the "themitical," referring to the normative, system-maintaining role of virtuous discourse and conduct (Faubion 2010: 95). One aim of this volume is to re-examine and retrieve from traditional Chinese religiosities, ontologies, and ethics, elements for restructuring our current anthropocentric values and constructing a new environmental ethics. This pursuit is in the spirit of what Deleuze and Guattari (1987) call "a line of flight" out of the world's current industrial-capitalist normative order and utilitarian ontology, to build new environmental ethics and ontologies for the Anthropocene age.

Michael Lambek associated the notion of "ethics" with such terms as "freedom, judgment, responsibility, dignity, self-fashioning, care, empathy, character, virtue, truth, reasoning, justice, and the good life for humanity" (Lambek 2010b: 6). Many of these terms came out of a European Enlightenment perspective focused on individual judgment and reason. As Jeffrey Nicolaisen in chapter 1 shows, even the Enlightenment notion of "human rights" has a deep Christian provenance. The work of constructing a diverse global environmental ethics cannot restrict itself to either the Enlightenment or Christian traditions. The present volume seeks to expand ethics inquiry in three ways: (1) broadening the anthropology of ethics from merely considering human interactions to addressing the ethical conduct between humans and nonhuman beings, divinities, objects, and forces of the natural world; (2) encompassing non-Western and non-Christian traditions; (3) building upon Lambek's (2010c) discussion of ritual as embodied ethics, inspired by the work of Roy Rappaport's theories of ritual (1999). The entanglement and mutual support of ethics and ritual performance was a recurrent theme upon which ancient Confucians had reflected upon and theorized extensively (Yang 1994: 216–244).

In the context of Chinese cultural history, Lambek's list of ethical terms would look quite different. The Chinese list might include such ethical notions as "obligation" (義務), "reciprocity" (報), "generosity" (慷慨), "charity" (慈善), "filiality" (孝), "kinship feeling" (親情), "friendship" (友情), "righteousness" (義氣), "respect" (尊嚴), "ritual and propriety" (禮儀), "penitence" (懺悔), "compassion" (慈悲), "virtue" (德), "reverence" (崇敬), and "sacrifice" (祭, 犧牲). For the sake of brevity, I cannot delve into how the semiotics and politics of these terms waxed and waned. The first nine expressions tend to focus on relations among humans. However, beginning with "ritual and propriety," the last six terms were deeply woven into Chinese religious cultures, and often invoked or involved human-nonhuman relations, with divinities, ancestors, animals, or gods. "Ritual and propriety" are key terms from Confucian ethics, where kinship ethics were expressed, embodied, learned, and mobilized through ritual performance involving ancestors. Similarly, Roy Rappaport (1999), following J. L. Austin, calls ritual practice "performatives" that harbored the ethical codes of the past, to which performers of rituals gave tacit acceptance without explicit belief or reasoning, due to their ritual participation. "Penitence" and "sin" (罪) were especially important in Daoist and Buddhist traditions, and there were many rituals of repentance to compensate for the loss of merit caused by sins against people, ancestors, gods, buddhas, and animals, who are reincarnations of past humans. "Compassion" was a potent Buddhist notion that enjoined people to empathize with and relieve the suffering of sentient beings that crossed the human-nonhuman divide. We will see how this Buddhist notion is put into action today in chapters 1, 2, and 6. "Reverence" was the proper attitude expected toward deities and ancestors, and the notion of "virtue" ranged from its ancient provenance of the moral authority of the just ruler in the ancient classic *Dao De Jing* 《道德經》, to the modern common term for "morality."

As the left-hand radical for "ox" (牛) in the Chinese ideograph for "sacrifice" (犧牲) reveals, the ancient Chinese saved the most treasured products of their labor, animals and meat, to offer to the ancestors or gods, who controlled the cosmos and could send down natural disasters. According to Michael Puett, from the Bronze Age (2000–771 BCE) to the Warring States (475–221 BCE) (and into late imperial China), there was the idea that "natural phenomena were governed by distinct, active deities" (Puett 2002: 96–97), as seen in this passage from the classic *The Book of Rites*:

> 燔柴於泰壇，祭天也；瘞埋于泰折，祭地也，用騂犢。埋少牢于泰昭，祭時也；相近于坎壇，祭寒暑也。王宮，祭日也；夜明，祭月也；幽宗，祭星也；雩宗，祭水旱也；四坎壇，祭四時也。山林、川谷、邱陵，能出雲為風雨，見怪物，皆曰神。有天下者，祭百神。
>
> —《禮記，祭法》

To sacrifice to Heaven, pile wood to burn the sacrificial victims on a high earth mound altar; to sacrifice to Earth, bury the sacrifice in the earth under the altar, and use a red calf. To sacrifice to Time, bury lambs under the Taizhao altar. To sacrifice to the gods of winter and summer, offer the sacrifice close to the altar in the earth pit. The "King's Palace" is the sacrifice to the Sun, the "Midnight Brightness" is the sacrifice to the Moon. The "Hidden Ritual" is the sacrifice to the stars, the "Summer Rain Sacrifice" is the sacrifice to request rain when there is a drought, and the "Four Altars in the Earth Pit" is the sacrifice to the gods of the four seasons. The mountains, forests, rivers, valleys, and hills that can send out clouds, make wind and rain, and cause to appear strange phenomena are called spirits (神 *shen*). He who possesses All Under Heaven sacrifices to the hundred spirits.[13]

—*The Book of Rites:* "The Method of Sacrifice" (Wang, M 1987)

In this passage, we see that in ancient China, political authority and the well-being of the human population were tied into, indeed dependent on the favor of nature spirits, forming a symbiotic relationship between the human and natural-divine realms. If humans took all the products of nature for themselves and failed to offer them to the nature spirits, or if the sovereign got distracted by his licentiousness, lost his virtue (德), and failed to keep up the sacrifices (Puett 2002: 301), they would incur the anger of natural-divine forces, bringing natural disasters. This ancient tradition of working to produce the goodwill of nature deities is today still alive in Daoist ritual practices of supplicating the pantheon of gods.

Heaven was the supreme nature deity, and the ancient Chinese did not anthropomorphize Heaven, nor represent this deity in human form. Thus, from ancient times to the 1911 Republican Revolution, the higher force that the ancient kings or emperors propitiated and answered to, was Heaven, a natural force. The Mandate of Heaven (天命) taught that the Son of Heaven (天子), the semi-divine sovereign, was selected by and served at the mercy of Heaven. Once the sovereign's ethical behavior and those of his officials deteriorated, Heaven would send down portents of natural disasters to indicate his removal of his Mandate, which would be awarded to another royal clan. Thus, the human polity was imagined as ethically answerable to higher natural forces (Miller 2017: 27–28). This greatly contrasts with modern times, when natural forces are no longer animated or imagined as being in charge of human destiny, but have become a series of inert trade items, like "real estate," "soft commodities" (pigs, soybeans, corn, palm oil), and "hard commodities" (minerals, rubber, coal, oil). Whereas the ancient Chinese attitude was one of awe, respect, and reverence, the natural world of modern times has lost its protective spirits and is now exposed and stripped, coming under the drills and bulldozers of modern technology. Modernity renders natural elements as inert, passive objects without consciousness, feelings, or the ability to fight back against grave lapses of human ethics (Merchant 1992; Descola 2013). The recent increase of extreme natural disasters due to climate change may return us to the past stance of awe and respect of natural forces. At the same time, we can never go back to ancient times, for we now have nearly eight billion humans to feed, clothe, and house, and there is barely any wildlife left to command our awe. Nevertheless, we can ask how certain elements of premodern environmental ethics might be adapted today to alleviate our ecological crisis and promote human protection and harmony with nonhuman life.

Lambek's edited volume on the anthropology of ethics also emphasizes human language as the key dimension of ethics, featuring the tools of linguistic anthropology as its approach to the study of ethics. However, when ethics are seen as "intrinsically linked to speaking as much as it is to action" (Lambek 2010: 5), then human language is privileged. This focus on human language restricts us to human ethics, shutting out consideration of the ethics of human-animal, human-plant, human-natural force, and human-divinity relations.[14] One certainly needs some form of communication in order to engender, nurture, and maintain the ethics of care, empathy, and obligation between living things. However, other animal species do not rely on human language or linguistic systems. As David Abram (2010) has shown in his poetic meditative prose, nonhuman animals employ other forms of communication, such as whole-body movements, gestures, dance steps, vocal sounds, eye movements, patterns of flight, emissions of body odors, and display of colors, feathers, or fur standing on end. Many of these animal communications are warnings directed at predators, but they are also cries for help, gestures of friendliness, or mischievous play to elicit responses from their own or other species.

Let us examine a Chinese mythological narrative that combines the elements of creation myth, nonverbal language, human-animal-divinity relations, environmental ethics, and ritual performance. From an ancient shamanistic culture dated to around 2200 BCE, we have the myth of sage-king Yu the Great (大禹), who subdued the great flood, and was appointed the first king of the Xia Dynasty (夏) (Yuan 2004). A great flood inundated the land, causing much suffering among the people, who lost their crops, fields, and homes and had to live in dank mountain caves, nest in trees, and compete for dry patches of land with ferocious wild animals. Yuan He's account (2004: 223–234) describes Yu as a shamanistic figure who was at times dragon, animal, divinity, and human. Yu was born as a dragon that emerged out of his father's corpse, having gestated there for three years, and later he transformed into a human, a bear, and a God. The Lord-on-High appointed Yu to continue his father's work of stopping the floods, and bequeathed Yu with a magical soil and dragons to assist him. While gazing down upon the Yellow River from a clifftop, Yu saw a tall thin man with a white face and a fish body who jumped out of the waves and presented him with a large green stone with an incised pattern on it, before disappearing back into the water. This creature was the River God (河伯), and his gift to Yu was a sacred river map to guide him in his flood control. Yu also relied on a black tortoise to carry the magical soil across the land and he used the tails of his dragons to carve out deep channels and riverbeds in the earth to guide the floodwaters to flow out across the land and into the Eastern Sea. The highest of the levees and dikes that he built up with the magical soil along the rivers later became China's sacred mountains. Much is made of how Yu's virtue of self-sacrifice and his love for the people: his hands were very calloused from his difficult labor and his toes were crushed and his feet twisted. Nor did he have time to visit his wife or child over many years, out of his sense of urgency to save the people.

The myth of Yu the Great features an example of nonverbal communication between human and animal and its ritualization. In Yu's travels through southern China's coastal area, he encountered problems with moving giant boulders and rocks. He saw how an exotic bird could do this through a ritual performance and "secret incantations" (禁咒) with its feet: the bird followed a series of strange footsteps and pacing that enabled it to move large rocks. Yu imitated the bird's left-right, left-right footsteps, and adopted the bird's rhythmic pacing, which gave him success in digging trenches and moving mountains to channel the water. In later ages, people called this pacing the "Steps of Yu" (禹步), and they came to be integrated into later Daoist rituals, also called "stepping across the Dipper and traversing the Dipper handle" (步罡踏斗) (Zhang Z 2000). In Daoist ritual, the priest performs these ritual steps as if he were on a cosmic journey stepping across the seven stars of the Northern Dipper (北斗).[15] Like a shaman's flight through the nine layers of the Heavens, and continuing the shaman's closeness with animals and birds, the Daoist priest dances the Steps of Yu like that bird, calling down the gods to enjoy the ritual and receive the priest's entreaties to help his clients. It is true that this ancient myth is about a great hydraulic engineering project in which the natural landscape is carved out to benefit the human population and its agricultural activities, thus founding China's earliest dynastic state, the Xia. However, the moral for our age is that there was no stark separation between humans and nature, but a great deal of human-animal-divinity transformation. Yu's human engineering of the natural environment is accomplished with the cooperation of dragons, bears, tortoises, a bird, the River God, and more. Yu's success in controlling the flood waters lies not only in persuading animals and divinities to help him, but also in embodying virtue and self-sacrifice to move the gods. The story shows that a benign natural environment conducive to human flourishing depends on the personal virtue of human heroes, good relations between human and other species, the ritual performance and incarnation of environmental ethics, and a reverence for the higher powers of natural/divine forces.

III. THE ONTOLOGICAL TURN AND CHINESE RELIGIOSITIES

Ontology is the study of *being*, *existence*, and *reality*, with which Western philosophy has long engaged, from Plato and Aristotle in ancient Greece to Descartes and Heidegger in the modern era. For the most part, ontology in the history of Western philosophy focused primarily on the *human* existential condition. The recent "ontological turn" in cultural anthropology (Kohn 2015; Holbraad and Pedersen 2017) broadens this field beyond the comparative study of human "cultures" or "systems of representation" to encompass the phenomenological and experiential aspects of both human and nonhuman (animals, plants, and divine beings) forms of being, life, existence, and agency. The new questions asked now include questions such as: In what ways do human consciousness and senses of the self, diverge from other forms of life? How can human life best interact with other life forms? Are there possibilities of interspecies communication, understanding, or cooperation? Before modern secularism and the mechanistic Newtonian universe, things in the natural world were alive and "animated," exercising agency, so that humans were not the main active agents. The recent ontological turn entails the questioning of modern Western rationalities of anthropocentrism, capitalist profit, consumerism, and nation-state economic and technological competition that have now become globalized and propel our ecological crisis. These modern rationalities all depend upon an ontology that renders other life forms and natural features inert, passive, and inanimate, to be acted upon by human agency, which is elevated above all.

Theorists of environmentalism have brought our attention to a key moment in modern Western thought, when an ontological separation and opposition of nature from culture emerged. Modernity, according to Bruno Latour (1993), was made possible by a process of "purification," the radical separation of nature from culture, the human from the nonhuman, the natural from the supernatural, religion from science, and a religious cosmos from the *real*, defined as this temporal life of material and social existence. Carolyn Merchant (1992) showed how the nature/culture opposition accompanied the emergence of a Newtonian inert mechanistic universe that replaced older vitalistic cosmological visions. This mechanistic universe became the object of the search for universal laws that could be encapsulated and predicted with mathematical precision because the contingencies of life forms and their random agency were removed or neutralized into inert "matter." Although much of modern science has since moved on from the Newtonian paradigm, it still drives most of humanity. For Philippe Descola in his book *Beyond Nature and Culture*, what enabled the modern Western ontology he calls "Naturalism" was the invention of linear perspective in European painting, which endowed the observer with a commanding gaze and modern Europeans with the occularcentric ability to conquer distance and space through mathematical equations (2013). Thus began the scientific search for universal laws of matter that humans share with other organic or inorganic objects, but human consciousness, reason, and culture are set aside as unique and different from other matter in the universe. What religious environmentalism and the exploration of alternative ontologies can contribute today is the re-animation or revitalization of the nonhuman world, so that humans can develop empathy for this world, and feel that it is part of us, and our futures are enmeshed and interconnected.

Recent questionings of modern ontology and its radical separation between nature and culture have produced an anthropological exploration of non-Western ontologies, or modes of being and experiences of reality among different species and life forms in the natural world. Chief among them are ethnographies of the Amazonian indigenous people in South America by scholars such as Eduardo Viveiros de Castro (2012; 2014), Philippe Descola (2013), and Eduardo Kohn (2013). Viveiros de Castro sought to reveal the limitations of modern "multiculturalist" cosmologies, which only compare human cultures. For him, Amerindian cultures do not think of themselves as mere "cultures," and he proposed the idea that ontologically, they instead think and act in terms of "multinaturalism." Multinaturalism posits that there is a spiritual/conscious unity and continuity between human beings, spirits, and animals, but there is corporeal diversity among them. Amerindians do not observe animality as a unified domain opposed to humanity, rather they treat nonhuman animals and spirits as "persons" who regard themselves as humans. More specifically, Amerindians practice "perspectivism," where these nonhuman "persons" (what we call jaguars, wild boar, or monkeys) even turn the tables, so that from their point of view, human beings look like, and are treated as "animals," "spirits," or even "prey" (2012: 83). Eduardo Kohn's fieldwork in Ecuador among the Runa people led him to challenge the "linguistic turn" of structuralism and poststructuralism, which privilege human language as the sole purveyor of signification and communication. He found that the Runa often deployed a nonlinguistic sign system that brought them closer to the ways that animals communicate, including signs, bodily movements, facial and limb gestures, and onomatopoetic utterances/words (2015). In this way, the Runa do not compartmentalize human versus animal versus spirit, but facilitate interspecies communication and feelings of shared culture with them. These indigenous ontologies thus provide alternatives to the modern nature/culture divide.

Philippe Descola (2013) took these alternative non-Western ontologies and classified them into four categories, based on how each ontology emphasized or compared the interiorities or outward physical features of different species or forms of life (table 0.1).

Animism is what Descola describes as ontologies like those of Amerindians, which find a common shared interiority across different animal species or life forms, while the external physicalities are regarded as dissimilar. *Totemism*, such as Australian aborigines' clan identity with an animal totem, or many forms of ancestor worship, takes both interiorities and physicalities to be similar and intimately connected. *Analogism* finds both interiorities and physicalities to be dissimilar but seeks to find order in the chaotic plenitude

Table 0.1   Philip Descola's classification of four basic ontologies (2013:122)

Similar interiorities Dissimilar physicalities	Animism	Totemism	Similar interiorities Dissimilar physicalities
Dissimilar interiorities Similar physicalities	Naturalism	Analogism	Similar interiorities Dissimilar physicalities

of multiplicitous forms of life through the construction of hierarchies of linked beings and species, or through narratives about the relatedness and analogies between microcosm and macrocosm. The medieval Christian European construction of the Great Chain of Being orders all earthly and celestial forms into a hierarchy of: God, angels, humans, animals, plants, and minerals. Finally, *Naturalism* is what Descola calls our modern Western ontology, which introduces a sharp break between nature and culture, and has diffused across the globe, with a vengeance in modern China. In the ontology of Naturalism, the interiorities of different species are regarded as dissimilar, while the basic substances of physicalities are believed to be similar across species or kind. In Naturalism, since humans do not share interiorities with other living beings, our ethics do not extend to them, nor do our obligations or empathies. Some may find Descola's highly structuralist schema[16] too neat and tidy, and even rigid. Others might even object to his unwitting replication of the age-old Western binarisms of the soul and the body, and its modern Cartesian reproductions of mind (interiorities) versus body (physicalities), or Hegelian idealism versus materialism. These binarisms were not emphasized in traditional Chinese cosmologies or ontologies. Still others may object to the lack of historicity and flexible flows in his structuralist ideal categories, and prefer a practice-oriented approach, in the manner of Pierre Bourdieu's (1977) critique of structuralism as focused on social rules, rather than examining the actual contingent implementation of rules and ideals (see section IV).

When we borrow Descola's schema and turn our gaze to traditional Chinese religiosities, we find a messy and riotous amalgamation of *all* three non-Western ontological categories. Chinese history is the gradual absorption of Neolithic and tribal societies by various archaic states, kingdoms, and empires, and what I have called "religious hierarchical encompassment" (Yang 2020: 55, 94, 274), the layering and absorption of older indigenous and tribal religiosities by later Axial Age, medieval, and late imperial traditions of Daoism, Confucianism, and Buddhism. Thus, it should not be surprising to find all of Descola's ontologies and more, operating in premodern China, as chapters 3 and 5 by Tristan Brown and Robert Campany show. The point here is to de-naturalize the modern Western ontology that has become globalized, in which there is a hard separation and difference between humans and nonhumans, and by extension, culture versus nature. Throughout China's three major religious traditions (Confucian, Buddhist, and Daoist) and popular religiosity, there is a great amount of interaction and transformation between humans and animals or spirits. Chinese shamanic cultures feature the transformation of humans into animals and gods. Most of the gods and immortals of popular religion and Daoism were once human beings who sacrificed themselves or made a contribution to others. Buddhas and Bodhisattvas were once human beings who cultivated themselves, achieved wisdom, and performed meritorious deeds. Ghosts and ancestors were all once human beings with limited mortal bodies and lifespans. For Confucians, sages and emperors were semi-divine and semi-human. Suffice to say, in premodern China, there were no sharply demarcated boundaries between the realms of humanity, nature, and the supernatural.

There are some pockets of China today where one can still find a living culture of *animism*, where tree, tiger, weasel, and fox spirits still roam, and receive offerings (Kang, XF 2005). Many a resident tree spirit have in traditional times saved the tree from being cut down, and fox spirits who lurk around temple grounds and forests have protected those sites from being destroyed. Furthermore, Chinese gods, Daoist immortals, Buddhas and Bodhisattvas, and ghosts often think and behave like human beings, but their ethereal forms, superhuman physical prowess, ability to perform miracles, to fly and be at many places at once are very different from humans. Indeed, in Chinese spirit medium cults, lacking human physical traits like speech, the gods must take over the bodies and tongues of human mediums in order to communicate with other humans. When nonhuman physical forms are endowed with humanlike minds and feelings, we have animistic ontological processes. These often express their very humanlike resentments and revenge for human disrespectful or sinful behavior. Ghosts, of course, have preserved many graveyards, because humans fear to violate their space.

As Robert Ford Campany's chapter 5 in this volume shows, in medieval China, animal and plant spirits as well as deities often communicated with people in their dreams, and people underwent transformations into them. Chinese dreaming did not strengthen what Charles Taylor (2011) called the post-Reformation "buffered self," whose boundaries were clearly demarcated and separated from other selves and other species. Chinese dreaming, writes Campany, "afforded a portal for face-to-face encounters with Others across ontological, taxonomic, spatial, and linguistic gaps." His account includes human interactions with a spirit, a tree, ants, as well as a human transformation into a fish who gets chopped up. In contrast to the modern Western pattern of understanding dreams only in terms of how they reveal the individual's innermost desires and the invisible workings of the self, Campany finds that Chinese dreamtime gave play to disparate tropes such as soul-splitting and wandering, seeing into the future, interactions with nonhuman entities, and exorcism. When neither dreaming nor dream interpretation re-enforces the stable human self, the seat of human desire, perhaps we are better able to avoid the possessive individualism of our modern culture of consumerism and accumulation.

Chinese ancestor worship comes close to *totemism*, for the descendants share the same interiority as well as physical substance with their ancestors, and the goal is to prolong and reproduce the spiritual and physical substance of the ancestors in genealogical continuity down through the ages. Whereas archaic Chinese Bronze Age shamanism may have worshiped animal spirits as clan totems (Chang 1983), in later ages, ancestors became humanlike. In Confucian ancestor worship, much care was devoted to the physical corpse, which should be kept whole and buried intact, as opposed to the Buddhists who cremated the dead. The corpse shares the same physical substance as its parents and its own children and descendants, while the interiority of the ancestor is also humanlike, so that it can understand and communicate with its descendants. The ancestors' bones interact with the flows of "primary breath" (氣 *qi*) flowing through the earth, to produce auspicious forces to benefit their descendants. Much attention was also devoted to the funerary ritual and the construction of the tomb, as expressions of filiality. The sacredness of ancestors meant that people down through the centuries tried to keep the burial areas quiet, clean, and peaceful, so that the ancestors may rest undisturbed. A hill or mountain dotted with tombs often provided shelter for trees, bushes, animals, and birds.

In chapter 4 by Liang Yongjia, we see how in 2012, even after a century of political attacks on the legacy of Confucius and the systematic dismantling of the venerable and complex Confucian education and ritual system, an insensitive local government campaign in rural Henan Province to dig up people's ancestral tombs still raised the hackles and anger of the people. The impetus for this local government campaign was economic: the search for scarce land to promote industrial development at all costs. In the state developmentalist ideology widely shared by officialdom across China, industry is much more prestigious and lucrative than agriculture or forestry. Liang examines the ensuing civil activism against this campaign, in which self-identified elite Confucianists across the country joined in an unintended alliance with the local rural folk. Together, they successfully put an end to the unpopular local government campaign of removing graves. The success of indigenous wisdom against a developmentalist ideology was due to a confluence of different agendas. For the elite educated Confucianists, their efforts were couched in terms of constructing a new Confucian cultural nationalism based on the ethic of "filial piety" to correct an overly "Westernized" secular state. However, the local rural folks' resistance was based on the concern to protect the good *fengshui* (風水) or the healthy flows of *qi* through the bones of the ancestors in the earth, that would bring blessings through the generations and ensure the perpetuation of the living lines of ancestral spirits.

We can also see *analogism* operating quite pervasively in China, since the gods of both Daoism and popular religion are arranged into complex hierarchies and pantheons of interlinked deities. The parallelism between the earthly human bureaucratic hierarchy of officials and the celestial bureaucracy of ranked gods and their honorific titles forms a "model for" and a "model of" human conduct, according to Emily Ahern (1981), in a microcosm and macrocosmic relationship. In Chinese Buddhism, after death, one's *karma* determines how a human being may be reborn into a sentient being in any of the Six Paths of transmigrating souls, which are arranged into a graduated ranking of past and future lives. The Six Paths are, in order of ranking: deity, *asura* (bellicose demigod), human, animal, hungry ghost, and hell denizen, the most miserable fate. Wei Dedong's chapter in this volume examines how the Chinese Buddhist ritual of "animal releasing" (放生 *fangsheng*) is a way for Chinese Buddhists to accept, reproduce, and reinforce this Buddhist analogist ontology on the one hand, and on the other hand, to assert their agency in seeking to improve their chances of being reborn into a better life. The ritual expresses the compassion taught by Buddhism for the sufferings of other sentient beings, through liberating a captured animal or fish that was destined for a human dinner table as food. At the same time, the chapter also reveals the shortcomings of ritual practice in today's commercial society and deviation from the original intent of compassion for animal suffering.

The ancient Chinese religio-environmental ethics of *fengshui* or "Chinese geomancy" presents an anomalous category that cannot easily fit into Descola's scheme of three non-Western premodern ontologies (animism, totemism, and analogism). Nevertheless, *fengshui* may have a large role to play in the development of new environmental ethics for modernity, alongside many Native American, Australian aboriginal, and other indigenous people's sentiments about sacred landscapes. The Chinese culture of *fengshui* shares many features with what John Grim calls "indigenous lifeways," religio-environmental expressions of "an intimate relatedness between . . . embodied self, the native society, the larger community of life in a [bio]region . . . or ecology, and the powerful cosmological beings . . . in ritual actions and mythic narratives" (2006: 288). For premodern Chinese and many different indigenous people around the world, the physical features of the landscape and bioregion in which they live are sacralized and embedded in larger cosmological orders. In turn, the sacralized landscape forms an intrinsic part of, or an integrative force in their community's social cohesiveness. The art of *fengshui* is based on facilitating the smooth flow of "primary breath" (*qi*), an invisible force that traces back to the origin of the cosmos. *Qi* flows through both the landscape and the bodies of living beings. The proper alignment of human constructions (like tombs, temples, or houses) and activities with *fengshui* patterns will bring protection, health, and good fortune to its residents, lineages, and communities. The determination of auspicious siting and orientation of human built forms is generally made by *fengshui* masters, wielding magnetic compasses which tap into electro-magnetic fields. Thus, in contrast to Descola's ontological category of modern Naturalism (more given further), the worldview of *fengshui*, like many indigenous sensibilities around the world, is of an animated, vitalistic, and sacred landscape whose flows and patterns must not be casually violated by human actions.

Tristan Brown's chapter 3 examines Qing Dynasty legal court cases, in which *fengshui* integrity and the protection of a local community's livelihood and cosmological well-being are commonly invoked to resist pressures for opening up the area for mining. To be clear, in Chinese cultural history, *fengshui* was never conceived or designed as environmental protection. However, the effect of invoking its protection was to regulate the commercial exploitation of the land and what lay aboveground and below. Brown shows how the ancient Chinese notion of "meridians" or "veins" (*mai* 脈) traversed what to us moderns are the ontologically different categories of the human body and the landscape/earth, channeling the flow of *qi* across different forms of life and nonlife, human and nonhuman elements. While *qi* flows through the meridians of the body, it also flows through the deep network of mineral and coal veins, and "dragon veins" (*longmai* 龍脈) (which determine political potency and good fortune for family and kinship lines) in the landscape. This movement of *qi* traversing human and inorganic entities in the cosmos is reminiscent of Zhang Zai's epigraph at the beginning of this chapter. These mineral veins must be protected, for their violation would destroy the health and well-being of the body, families, kinship groups, and local community. When the imperial state responded positively to plaintiff pleas for protection of *fengshui*, they allowed kinship groups and local communities to protect their livelihoods and the ability of cosmological forces flowing through the land to protect them. Thus, while late imperial China did not conceive of *fengshui* in terms of what we now call "environmental protection" or resistance to "economic development," appealing to *fengshui* protection did sometimes achieve just that. In the eighteenth and nineteenth centuries, as commercial and demographic forces exerted increasing pressures in China to open up more mines to extract coal and minerals, the imperial state acted as a check on "development," by considering whether the *fengshui* of local communities would be damaged.

Given the globalization of the European ontology that Descola calls modern "Naturalism," any attempt today to construct a new religious and environmental ontology must take into account, build upon, or seek to overcome this originally Western ontology. Jeff Nicolaisen's chapter 1 in this volume traces the Christian anthropocentric provenance of "human rights" discourse. For John Locke, the equality of humans was based on their God-given capacity to reason that God exists and created the universe. This spark of the divine is exclusive to humans, and sets humans apart from nature. Thus, environmentalist movements today that are guided by rights discourse stem from conscious or unconscious notions of the God-given "human right" to a healthy environment. Nicolaisen examines how the Taiwanese Buddhist nun

Shih Chao-hwei (釋昭慧) is attempting to forge a new non-anthropocentric global ontology. Master Chao-hwei takes the European Enlightenment discourse of "human rights" that has penetrated the United Nations, and infuses it with a Buddhist perspective. In her campaigns for animal rights, Master Chao-hwei introduces the more inclusive Buddhist concepts of the "equality of life" (*shengming pingdeng* 生命平等) and "protecting life" (*husheng* 護生), that call for the protection of both nonhuman and human lives. For Master Chao-hwei, equality derives not from rationality or a divine spark, but from "dependent-arising"—the fact that all life arises from causes and conditions. For her, suffering, not the capacity to reason, is the standard for ethical consideration. With no distinction between humans and other forms of life, there is no room for a concept of nature distinct from human life. This chapter deploys the concept of the rights of nature to re-examine Chao-hwei's writing and evaluate how the legal rights of nature may be applied in an ontological system that does not include nature.

Environmental scholars and activists have also addressed the human relationship with inorganic matter and nonliving entities, such as rocks, rivers, and plastic objects. Indeed, Katherine Yusoff has called for the "remineralization of humanity," since our human pasts are preserved as fossilized calcium (bones) that are found in deeper geological layers. She also calls for granting agency to fossil fuels, products created out of the remains of once living plants and creatures from the Carboniferous Age, since they actively shape and constrain our contemporary life in the Anthropocene (Yusoff 2013). In answer to modern Naturalism, which created an inert and mechanical universe that operated like the mechanical springs of a clock according to universal laws of physics, some environmentalists have attempted to "re-animate" what modernity defines as the nonliving parts of nature. After decades of state inaction in stemming the terrible pollution of Hinduism's most important sacred river, that is also the deity Mother Ganga, environmental lawyers and Hindu religious forces joined together to reanimate the river in legal terms, so as to justify its right to environmental protection. In 2017, the court of the Indian state of Uttarakhand declared the Rivers Ganges and Yamuna to be "legal and living entities having the status of a legal person with all corresponding rights, duties and liabilities."[17] Mayfair Yang and Huang Weishan's chapter 2 in this volume also recounts how the Taiwanese lay Buddhist charity movement Tzu Chi Merit Foundation has sought to animate *things*, in order to instill the lesson that inanimate objects can be given new lives and usages in new forms. With the newly coined Buddhist notion of the "life of things" (*wuming* 物命), Tzu Chi devotees learn to anthropomorphize and personify inanimate synthetic objects, like everyday plastic waste, in order to imbue them with value and recycle them into new uses and future lives.

IV. FROM TEXTS AND TEACHINGS TO "PRACTICE": HISTORY AND CONTINGENCY IN RELIGIOUS ENVIRONMENTALISM

The ancient religious cultures that twentieth-century Chinese impatiently rejected, now hold the promise of helping to tackle ontological questions of modernity to reduce further environmental damage. They offer a vast repository of ancient environmental ethics that we moderns can tap into. Most of the early writings on Asian religious ecology, such as the pathbreaking Harvard University series: *Buddhism and Ecology* (Tucker et al. 1997); *Confucianism and Ecology* (Tucker et al. 1998; Tu 2001); and *Daoism and Ecology* (Girardot et al. 2001), and more recently, James Miller's delightful *China's Green Religion* (2017), focused on the explication of primary textual sources. These English language publications have also helped to inspire comparable textual investigations among scholars in China, some of whom (Chen 2010) have read the ancient Chinese sources through the "deep ecology" lens of Arne Naess, whose work has been translated into Chinese. In these efforts, ancient historical texts, medieval religious scriptures, and philosophical classics, are carefully mined for passages relevant to addressing our contemporary environmental problems. While these textual excavations have proven extremely enriching, some have also pointed out that sometimes, there is a tendency to romanticize past religious cultures.

When a culture or minority is nonhegemonic, weak, or declining, as ancient or medieval Chinese religious cultures are in today's secularized, hyper-commercialized, and urban technological Chinese society, there is a natural tendency to withhold critique and emphasize the positive about them. They are the non-threatening exotic "Other" whose loss we mourn, and from whom we wish to retrieve inspirations for the present. However, we must also remember that these religious traditions were deeply embedded in their own hegemonic power systems with which they were at times complicit, and at other times questioning or defiant. Most importantly, they evolved over centuries and millennia when human beings were much more concerned about their physical survival, and vulnerable to the natural calamities of floods, insect plagues, droughts, famines, and predatory wild mammals. When preindustrial technologies had not yet threatened the environment like today, these ancient religious traditions were not urgently propelled by a need to protect the nonhuman environment. Given that natural forces for the ancient Chinese must have often seemed frightening and overpowering, some of their actual practices may not have been so environmentally friendly, even as many of the writings they left behind offer us a different ontology and inspire us today. In other words, since religious cultures in the past were not specifically designed to protect the environment, we must grant them a great deal of independence in our expectations. Many dimensions of the past cannot address the particular concerns we have today. The search for environmental ethics and ontologies in traditional Chinese religiosities and spiritualities must acknowledge the contradictory features and the not always benign dimensions of religious cultures.

Despite the non-anthropocentric ontologies, cosmologies and ethics of premodern China discussed earlier, we also have much evidence of environmental damage throughout China's past, as examined in Robert Marks' excellent volume, *China, an Environmental History* (2017). The Chinese landscape was once abundant with large mammals, such as elephants, rhinoceroses, tigers, bears, deer, and monkeys, species which, even before modernity, became extinct or endangered. At the ancient Bronze Age archaeological sites of Sanxingdui and Jinsha outside the modern city of Chengdu in Sichuan Province, Chinese archaeologists dug up amazing piles of giant elephant tusks dating to about four thousand years ago. Sacrificed to the ancient gods, these precious objects revealed the abundance of huge elephants who used to roam the Chengdu plains, but according to Mark Elvin's magisterial book, *The Retreat of the Elephants* (2004), had almost disappeared from China by the Ming Dynasty (1368–1644 CE). Over the course of several millennia, human population growth and activities such as agriculture, irrigation, deforestation, warfare, using wild animals as food and medicine, and the damming and manipulation of the course of river flows have altered the landscape and depleted its forests and wildlife. In pursuit of state activities such as warfare and the building of infrastructure, palaces, and ships, great primeval forests were cut down. These premodern environmental damages raise many questions about the actual practice of religious doctrines in social life, and the strength of religious institutions: How persuasive were Buddhist injunctions against killing sentient beings? How common was the implementation of Daoist teachings of governance through "non-intervention" (*wuwei* 無為) and following the natural rhythms and forces of the cosmos? How much authority and influence did religious institutions have against other institutions that promoted agriculture, human population growth, and warfare?

When we juxtapose the eloquent sentiments of environmental ethics in China's religious and spiritual past, with accounts of premodern environmental destruction, we recognize that religious texts espouse many honorable intentions, beautiful ideals, and inspiring narratives, but they are not always implemented, or they are selectively applied in social practice. Indeed, social practices may even belie or even defy the stated religious ideals that are found in written scriptures or philosophical texts. At the same time, eloquent religious scriptural teachings are often most vividly expressed through the seemingly most inconsequential acts and behaviors of everyday life.

We can avoid over-romanticizing Chinese religious texts by examining how religious ideals were or were not actually put into "practice," and how changing historical and environmental conditions often led to new interpretations and practices. As Pierre Bourdieu, the most well-known practice theorist, has argued, official discourses and dominant structures of power are only as good as the degree to which agents on the ground put them into practice. The implementation of religious teachings is often imperfect and selective. Religious structures, institutions, and traditions can only instill in people deep dispositions, but they cannot guarantee that agents will always act out or realize these dispositions. That is because *habitus*, those cognitive, bodily, and motivating structures that human agents absorb in the course of socialization in a culture (1977: 76) are subject to vicissitudes of history and different contingencies. In putting religious doctrines into practice, agents do not merely obey religious *rules, regulations, and norms*: they consider the particular situation they are in, they *strategize* their sequence of actions, they *improvise* with what the situation presents, and they *time* their actions and speech.

Past environmental destructions took millennia and centuries to accomplish. Compared with modern times, they pale in the scope and speed of human impact on the environment. With modernity, we have increasing secularism and the decline of religious ontologies and cosmologies, and in their place, arise the twin scourges of capitalist commodification of nature and nationalist developmentalism. In modern China, in less than a century since embracing modern Western linear social evolutionism, the discourse of "progress," and the opposition between nature and culture, China has more than caught up with the West in the objectification and utilitarian treatment of the natural environment.

Four chapters in this volume address environmental ethics in action in contemporary China. Since they are based on fieldwork in China, they all examine how, when religious environmental ethics and doctrines are actually put into "practice," the process is often problematic. In contemporary China, religious institutions are still constrained by political interests, and religio-environmental practices must always take into account the presence of the modern state and its officials, and the shifting political winds and economic conditions. The outcome cannot be assumed, and is far from predictable.

Wei Dedong's chapter 6 examines how the revival of the ancient Chinese Buddhist ritual called "releasing animals" or "releasing life" (*fangsheng*) has ironically had adverse impacts on the natural environment. This occurs when people eagerly rush to release captured fish and other species to satisfy the need to accumulate merit, but do not consider the science that would ensure the animals' or birds' actual chances of survival in the environment to which they are released. Indeed, in contemporary Chinese Buddhist culture, some people may have lost touch with the empathy for the suffering of the animals, and are more concerned with the accumulation of merit. Counter to its original religious goals of ritualizing Buddhist sentiments of "compassion" (慈悲) for other sentient beings, today in China, releasing animals has grown into a profitable industry with commercial links throughout the country. Ironically, innumerable animals die as a result of this industry, and the ecological damage caused in the process is incalculable. Wei examines one innovative transnational cooperative effort for releasing animals, which preserves the Buddhist tradition. He describes the unique cooperation between the Grace Gratitude Buddhist Temple in Chinatown, New York City, and two American secular animal rescue operations, the New York City Turtle and Tortoise Society and the Wild Bird Fund in Manhattan. In this program, Chinese Buddhists in the New York City area generate merit by helping to collect and heal injured turtles, birds, and other wildlife and make financial donations to these animal rescue agencies. This new and improved releasing method has integrated Buddhist thought with animal science, and has been reported in the media back in China.

The chapters by Adam Chau and Yang Der-Ruey both push us to wonder how much environmentalist efforts made by religious institutions in contemporary China are actually due to religious environmental ethics? In the present context of a highly secularized Chinese society and the continued struggle by many religious organizations to win the approval of secular state authorities who are wary of religion's power for mass mobilization, one must wonder about the extent to which older environmental ethics can still exert influence. Chapters 7 and 8 both reveal the importance of what Chau aptly calls the "politics of legitimation," as a primary motivating factor today in pushing Daoist or deity temples to adopt environmentally friendly projects. Daoist environmental ethics may be a rather dim factor in their decisions. In a state-secularized context where religious organizations still face a state which is nervous and sometimes hostile toward religious life, these organizations must often overcome their labels of "feudal superstition" or official fear of religious rebellions. To ensure their survival, helping the state to engage in environmental activities may provide the path to legitimacy for these temples. Adam Chau's chapter examines a number of distinctive "regimes of tree-mindedness." These obsessions about trees and their associated practices are collective, and are products of systematic, long-term cultural inculcation or large-scale sociopolitical mobilization. Each regime of tree-mindedness evolves over time, and different regimes encounter one another, borrow elements from one another, compete, and intermesh with one another. The Black Dragon King Temple in Shaanbei, north-central China, initiated the Longwanggou Hilly-Land Arboretum reforestation project in the 1980s upon the advice of a local government forestry engineer as part of the temple's legitimation politics. The project was lauded as a folk environmentalist initiative and it attracted urban-based environmentalist NGOs that were eager to find places for its members to plant trees. These NGOs, in turn, inherited practices from another tree-minded regime, the work-unit-organized tree-planting activities from the Maoist high socialist era. This case study shows the ways in which the global environmentalist regime of tree-mindedness and the Chinese state's fetishization of tree-planting ("the greenification of the Motherland") intermeshed with a local temple cult's tree-mindedness. The chapter includes a reflection on the role that local cults can play in carrying out environmentalism because of their commitment to their native locale, a locale-centrism that resonates with the state's China-centered cosmic-governance.

Whereas the Black Dragon King Temple's reforestation project was a resounding success, the chapter by Yang Der-Ruey on the Ecological Forest of Daoism in contemporary Minqin County, Gansu Province, reports on the *failure* of a Daoist environmental initiative. Writing against the grain of much intellectual promotion of Daoism as one of the most environmentalist religious traditions in the world (Girardot et al. 2001; Miller 2017), Yang's chapter presents a sad experience of how a Daoist environmentalist effort to deal with rapid desertification was doomed and thwarted in actual practice. Yang adopts a rare history of the *longue durée* in examining centuries of changes in human demography, Han Chinese settlement, Daoist religious and ritual activities, modern state initiatives, and environmental degradation. He shows how the concerted state secularization drives of the twentieth century led to the loss of Daoism's public role and its religious authority to undertake public relief efforts in Minqin County. This loss included the rainmaking ritual that local Daoist clerics used to conduct, which unified the community, linked moral virtues with good fortune and livelihoods, and strengthened the community's resolve to deal with natural disasters and combat the desert. When in 1996, local officials got the green light from the central state to counter desertification in the area, and they subcontracted the job to the local Daoists, Daoist leaders seized this precious opportunity to reinstate their religious authority and rebuild their Daoist organization and temple. Sadly, their hard work was not rewarded, and their efforts were aborted when the local river suddenly dried up. It was perhaps a case of anti-desertification efforts that were "too little, too late." The sudden depletion of the river caused the central state authorities to swoop back in, bringing in earth science experts and instituting big-budget state projects of population displacement and technological transformations of agriculture, overriding both secular and Daoist local efforts. The reader is left with the question: If the local Daoist self-governing organization had not been dismantled in the 1950s when the local population size and water capacity were still manageable, would local Daoists have been able to assert their religious authority to limit population growth and the destructive impact of agriculture?

Last, but not least, Mayfair Yang and Huang Weishan's chapter 2 explores how the Taiwanese Buddhist Tzu Chi Merit Foundation negotiates its religio-environmentalist activities in the context of a largely secular Shanghai society today, creating social niches of community voluntarism and grassroots education for moral reform. Tzu Chi environmentalism embeds Buddhist principles and moralities within a secular language. Its lay practitioners engage in activities that straddle the religious-secular divide: recycling waste products, vegetarianism and "ethical eating" of simple local foods, street-cleaning, transforming fruit waste enzymes into household detergents, and making eco-textile products from recycled plastics. With the newly coined Buddhist notion of "life of things" (物命 *wuming*), Tzu Chi devotees learn to anthropomorphize and personify inanimate synthetic objects, like everyday plastic waste, in order to imbue them with value and recycle them into new uses and future lives. These reformed Buddhist teachings have stressed the value of religious tradition and made ethical discourse and action again relevant in their attempt to break away from the capitalist system, Tzu Chi followers engage in challenging the social norms of *economic value*, but not the political system or its policies. These grassroots efforts were undertaken at the level of local neighborhoods, called "Dharma fields" (*shequ futian* 社區福田). Thus, Tzu Chi contributes to pushing the larger Shanghai trend of shifting from a work unit-based society to a neighborhood-based urban life, representing a religiously inspired civil society.

V. CONCLUSION

The operant terms in this volume are "Chinese religions," "the Anthropocene," "environmental ethics," "religious ontology," and "practice." The chapters in this volume address the three main religious traditions of China: Daoism, Confucianism, and Chinese Buddhism, as well as popular religion and *fengshui* or Chinese geomancy. The contributors write within a historical context of growing awareness and concern for life crossing a new threshold into the epoch of the Anthropocene, when humans are the dominant force in altering climactic conditions, the quality of the air we breathe, the landscapes we inhabit, the flows and quality of our water sources, and the mass extinction of other species. Even as we embrace the label Anthropocene to describe our era, we are also cognizant that now more than ever, we need to move away from the Western ontology that centers and elevates *Anthropos* ("Man") as the primary agency on the planet. Many of the chapters here explore traditional Chinese religiosities for alternative ontologies that recognize and animate the agencies of other life forms, for environmental ethics that extend care and compassion to nonhuman forms of life, and for non-anthropocentric rituals that perform the possibilities for coexistence with other species. The volume seeks to balance attention to the religious ideals and teachings found in the ancient and modern religious textual records, with the actual practices that enact or realize these teachings. The contributors realize both the great environmental potentials of traditional Chinese religious ontologies and ethics, as well as the many constraints, obstacles, and difficulties in operationalizing them in the modern secularist developmentalism of contemporary society.

NOTES

1. For a history of the modern state suppression of Chinese religions and secularization drives, see David Palmer and Vincent Goossaert (2011) and Mayfair Yang (2008, 2011).

2. For a very useful introduction to the still relatively new field of ecology and religion or religious environmentalism, see the book *Ecology and Religion* by John Grim and Mary Evelyn Tucker (2014), who are pioneers in this field. For a very good collection of essays on ancient and modern Chinese eco-religious practices, covering both Han and ethnic minorities, see *Religion and Ecological Sustainability in China*, edited by James Miller, et al. (2014).

3. This 1969 oil spill in Santa Barbara has since been superceded in size by the Japanese tanker Wakashio oil spill in the pristine waters of Mauritius in 2020, Shell Oil's Deepwater Horizon spill in the Gulf of Mexico in 2010, the Exxon Valdez spill in Prince William Sound, Alaska, in 1989. For information on the First Earth Day festival, see https://www.cecsb.org/the-history-of-earth-day/. Accessed Sept. 8, 2020.

4. See the *New York Times* list of Trump Administration rollback of about one hundred environmental rules: https://www.nytimes.com/interactive/2020/climate/trump-environment-rollbacks.html. Accessed July 22, 2020.

5. Cowan, Jill and M.T. McDermott, "California Today." *New York Times*, Sept. 9, 202. https://mail.google.com/mail/u/0/?tab=wm#inbox/FMfcgxwJXpNbVgVBmVWwdsHpcsKmwpss. Accessed Sept. 9, 2020; Krishnakumar, Priya and Swetha Kannan. "The Worst Fire Season Ever. Again." *Los Angeles Times,* Sept. 15, 2020. https://www.latimes.com/projects/california-fires-damage-climate-change-analysis/?utm_source=sfmc_100035609&utm_medium=email&utm_campaign=31196+Today%27s+Headlines+9%2f16%2f20&utm_term=https%3a%2f%2f. www.latimes.com%2fprojects%2fcalifornia-fires-damage-climate-change-analysis%2f&utm_id=13925&sfmc_id=567657. Accessed Sept. 16, 2020. Associated Press. "Epic Scale of Western Wildfires Grows as a Single Fire surpasses One Million Acres." *Politico*, Oct. 5, 2020. https://www.politico.com/news/2020/10/05/california-wildfires-grow-426523. Accessed Oct. 6, 2020.

6. The Chinese character 天 in this phrase, which for premodern China is usually translated as "Heaven" (supreme deity who is the father of the Emperor), is here translated as "Nature." The rationale for this translation is that the religious monarchical system of the Qing Dynasty had been deposed in the 1911 Republican Revolution, and by the Maoist era, Chinese society had become thoroughly secular, with few people believing in the Supreme Deity Heaven.

7. Su Lin-tan. "China's Carbon Neutral Push Gathers Pace..." in *South China Morning Post.* Feb. 20, 2021. https://www.scmp.com/economy/china-economy/article/3122419/chinas-carbon-neutral-push-gathers-pace-coal-fired-power?utm_medium=email&utm_source=mailchimp&utm_campaign=enlz-gme_trade_war&utm_content=20210223&tpcc=enlz-us_china_trade_war&MCUID=6aafd4a49c&MCCampaignID=3e86ced18a&MCAccountID=7b1e9e7f8075914aba9cff17f&tc=17.

8. Collyns, Dan. "'They Just Pull up Everything!' Chinese Fleet Raises Fears for Galapagos Sea Life." *The Guardian.* Aug. 6, 2020.

9. Carrington, Damian. "Pandemics result from Destruction of Nature, say UN and WHO" in *The Guardian*, June 17, 2020; Benatar, David. "Our Cruel Treatment of Animals Led to the Coronavirus." *New York Times,* May 24, 2020; Nyabiage, Jevans. "Human Activity Raises Risk of More Pandemics like Covid-19, warns UN Report." *South China Morning Post,* July 6, 2020.

10. Kamble, Karan. "China Faces Wrath of Rain—Fears Over Breach and Deformation of Controversial Three Gorges Dam." *Swarajya,* July 25, 2020; Chen, Frank. "Three Gorges Dam Weathers the Flood Challenge." *Asia Times,* July 30, 2020; Alice Yan. "China on Alert for Yangtze River Flooding as Storms Close In." *South China Morning Post,* Aug. 17, 2020.

11. *Taiwan News* (7/23/2020) quoted from Xinhua News of July 18, 2020. https://www.taiwannews.com.tw/en/news/3972343. However, a search of the latter source on July 29, 2020, found that the original Xinhua quotation had been removed. In a Xinhua News interview on July 21, 2020, a manager of the Three Gorges Conglomerate strenuously denied any buckling of the dam: https://china.huanqiu.com/article/3z8hflumVeo. Accessed Aug. 8, 2020. A Mainland Chinese news source, *Caijing Lengyan* (財經冷眼), put out a video of the hypothetical havoc and damage that the collapse of the Three Gorges Dam might wreak to populations downriver. This video was shared on Twitter: https://www.taiwannews.com.tw/en/news/3973169. Accessed Aug. 3, 2020.

12. Out of thirty-three voting members of the Anthropocene Working Group, twenty-nine voted on May 21, 2019, to accept the Anthropocene as a "formal chronostratigraphic unit," starting in the mid-twentieth century C.E. http://quaternary.stratigraphy.org/working-groups/anthropocene/. Accessed Jan. 20, 2021.

13. In my translation of this passage, I adopt Michael Puett's translation of the last two lines of the Chinese, beginning with "The mountains, forests, rivers . . ." and I have also consulted Wang Meng-ou's (Book of Rites 1987: 738) translation into modern Chinese and his annotations.

14. Although Lambek does recognize that "a sense of ethics that precedes, exceeds, or escapes specific words and acts" (Lambek 2010: 6) can exist, he does not really develop this alternative line of inquiry outside of his discussion of ritual as a performative utterance, following the stimulating work of Roy Rappaport (1999).

15. What the Chinese call "Northern Dipper" is the same as the Big Dipper in Western cultures, a constellation made up of seven stars.

16. After all, Descola was a student of Claude Levi-Strauss, who believed in the universal binary character of human thought, and asserted that they formed enduring cultural structures of thought.

17. "Ganges and Yamuna Rivers Granted Same Legal Rights as Human Beings." *The Guardian,* March 21, 2020. https://www.theguardian.com/world/2017/mar /21/ganges-and-yamuna-rivers-granted-same-legal-rights-as-human-beings#:~:text =The%20Ganges%20river%2C%20considered%20sacred,same%20legal%20rights %20as%20people.&text=The%20decision%2C%20which%20was%20welcomed ,equivalent%20to%20harming%20a%20person. Accessed Aug. 24, 2020.

ENGLISH AND WESTERN LANGUAGE BIBLIOGRAPHY

Ahern, Emily. (1981). *Chinese Ritual and Politics.* Cambridge: Cambridge University Press.
Bourdieu, Pierre. (1977). *Outline of a Theory of Practice.* Translated by Richard Nice. Cambridge: Cambridge University Press.
Chakrabarty, Dipesh. (2009). "The Climate of History: Four Theses." *Critical Inquiry*, no. 35, Winter, pp. 197–222.
Chakrabarty, Dipesh. (2019). "The Planet: An Emerging Humanist Category" in *Critical Inquiry*, no. 46, no. 1, pp. 1–31.
Clements, John M., Aaron M. McCright, and Chenyang Xiao. (2014). "Green Christians? An Empirical Examination of Environmental Concern Within the U.S. General Public." *Organization and Environment*, vol. 27, no. 1, p. 85–102.
Crutzen, Paul. (2002). "Geology of Mankind: the Anthropocene." *Nature,* vol. 415, January 31, p. 23.
Deleuze, Gilles and Felix Guattari. (1987). *A Thousand Plateaus: Capitalism and Schizophrenia.* Brian Massumi, trans. Minneapolis: University of Minnesota Press.
Descola, Philippe. (2013). *Beyond Nature and Culture.* Janet Lloyd, trans. Chicago: University of Chicago Press.
Duara, Prasenjit. (2015). *The Crisis of Global Modernity: Asian Traditions and a Sustainable Future.* Cambridge: Cambridge University Press.
Elvin, Mark. 2006. *The Retreat of the Elephants*: *An Environmental History of China.* New Haven: Yale University Press.
Faubion, James. (2010). "From the Ethical to the Themitical (and Back): Groundwork for an Anthropology of Ethics" in *Ordinary Ethics: Anthropology, Language, and Action.* Michael Lambek, ed. New York: Fordham University Press.
Finamore, Barbara. (2018). *Will China Save the Planet?* Cambridge: Polity Press.
Frohlich, Thomas C. and Liz Blossom. (2019). "These Countries Produce the Most CO2 Emissions." *USA Today.* July 14. https://www.usatoday.com/story/money /2019/07/14/china-us-countries-that-produce-the-most-co-2-emissions/39548763/. Accessed July 28, 2020.
Girardot, N.J., James Miller, and Xiaogan Liu, eds. (2001).*Daoism and Ecology: Ways Within a Cosmic Landscape.* Cambridge: Center for the Study of World Religions, Harvard University.
Goossaert, Vincent and David A. Palmer. (2011). *The Religious Question in Modern China.* Chicago: University of Chicago Press.

Grim, John A. (2006). "Indigenous Traditions: Religion and Ecology" in *The Oxford Handbook of Religion and Ecology*. Roger S. Gottlieb, ed. Oxford: Oxford University Press.

Jenkins, Willis. (2018). "Naturalized: White Settler Christianity and the Silence of Earth in Political Theology" in *Political Theology*, Oct. 1. https://politicaltheology .com/naturalized-white-settler-christianity-and-the-silence-of-earth-in-political-the ology/. Accessed Nov. 9, 2020.

Kang, Xiaofei. (2005). *The Cult of the Fox: Power, Gender, and Popular Religion in Late Imperial and Modern China*. New York: Columbia University Press.

Kohn, Eduardo. (2015). "Anthropology of Ontologies." *Annual Review of Anthropology*, vol. 44, Aug. 13, pp. 311–327.

Lambek, Michael, ed. (2010a). *Ordinary Ethics: Anthropology, Language, and Action*. New York: Fordham University Press.

Lambek, Michael. (2010b). "Introduction" in *Ordinary Ethics: Anthropology, Language, and Action*. Michael Lambek, ed. New York: Fordham University Press.

Lambek, Michael. (2010c). "Towards an Ethics of the Act" in *Ordinary Ethics: Anthropology, Language, and Action*. Michael Lambek, ed. New York: Fordham University Press.

Latour, Bruno. (1993). *We have Never Been Modern*. Catherine Porter, trans. Cambridge: Harvard University Press.

Marks, Robert B. (2017). *China: An Environmental History*. 2nd edition. Lanham: Rowan & Littlefield.

Merchant, Carolyn. (1992). *Radical Ecology: The Search for a Livable World*. New York: Routledge.

Miller, James, Dan Smyer Yu, and Peter van der Veer, eds. (2014). *Religion and Ecological Sustainability in China*. New York: Routledge.

Miller, James. (2017). *China's Green Religion: Daoism and the Quest for a Sustainable Future*. New York: Columbia University Press.

Naess, Arne. (1989). *Ecology, Community, and Lifestyle*. David Rothenberg, trans and editor. Cambridge: Cambridge University Press.

Puett, Michael J. (2002). *To Become a God: Cosmology, Sacrifice, and Self-Divinization in Early China*. Cambridge: Harvard University Press.

Rappaport, Roy. (1999). *Ritual and Religion in the Making of Humanity*. Cambridge: Cambridge University Press.

Schwadel, Philip and Erik Johnson. (2017). "The Religious and Political Origins of Evangelical Protestants' Opposition to Environmental Spending." *Journal for the Scientific Study of Religion*, vol. 56, no. 1, pp. 179–198.

Shapiro, Judith. (2001). *Mao's War Against Nature: Politics and the Environment in Revolutionary China*. New York: Cambridge University Press.

Shapiro, Judith. (2016). *China's Environmental Challenges*. 2nd edition. Cambridge: Polity Press.

Smith, William. (2020). *China's Engine of Environmental Collapse*. London: Pluto Press.

Subramanian, Meera. (2019)."Anthropocene Now: Influential Panel Votes to Recognize Earth's New Epoch." *Nature*, May 21. https://www.nature.com/articles/d41 586-019-01641-5. Accessed Sept. 3, 2020.

Tsing, Anna, Heaterh Swanson, Elaine Gan, Nils Bubandt, eds. (2017). *Arts of Living on a Damaged Planet*. Minneapolis: University of Minnesota Press.

Tu, Weiming. (2001). "The Ecological Turn in New Confucian Humanism: Implications for China and the World." *Daedalus*, vol. 130, no. 4, Fall, pp. 243–262.

Viveiros de Castro, Eduardo. (2012). "Cosmological Perspectivism in Amazonia and Elsewhere." (Four Lectures Given in the Department of Anthropology, University of Cambridge, 1998). Manchester: HAU Masterclass Series 1.

Viveiros de Castro, Eduardo. (2014). "Who is Afraid of the Ontological Wolf? Some Comments on an Ongoing Anthropological Debate." Cambridge University SAS Annual Marilyn Strathern Lecture, 30 May.

Yang, Mayfair. (1994). *Gifts, Favors, and Banquets: The Art of Social Relationships in China*. Ithaca, NY: Cornell University Press.

Yang, Mayfair. (2008). "Introduction." *Chinese Religiosities: Afflictions of Modernity and State Formation*. Mayfair Yang, ed. Berkeley: University of California Press.

Yang, Mayfair. (2011). "Postcoloniality and Religiosity in Modern China: The Disenchantments of Sovereignty." *Theory, Culture and Society*. vol. 28, no. 2, pp. 3–45.

Yang, Mayfair. (2020). *Re-enchanting Modernity: Ritual Economy and Society in Wenzhou, China*. Durham: Duke University Press.

Yusoff, Kathryn. (2013). "Geologic Life: Prehistory, Climate, Futures in the Anthropocene." *Environment and Planning D: Society and Space*, vol. 31, pp. 779–795.

CHINESE LANGUAGE BIBLIOGRAPHY

Chen, Xia 陳霞. (2010)《道教生態思想研究》(*Studies in Daoist Ecological Thought*). 成都: 巴蜀書社 Chengdu: Bashu Book Society.

《禮記今註今譯》(*The Book of Rites: New Annotations and Translation*). 1987. 王夢鷗註譯 translated & annotated by Wang Meng-ou. 臺北：台灣商務印書館.

Yuan, He 袁珂. (2004).《中國古代神話》(*Ancient Chinese Mythology*). 北京：華夏出版社.

Zai, Zhang 張載. (1999). "The Western Inscription"《西銘》in *Sources of Chinese Tradition: From Earliest Times to 1600*. Edited by W. Theodore De Bary and Irene Bloom. New York: Columbia University Press, pp. 682–684.

Zhang, Zehong 張澤洪. (2000). "論道教的步罡踏斗" ("On the Daoist Ritual Pacing of Stepping Across the Northern Dipper" in《中國道教》(*Chinese Daoism*), no. 4, pp. 7–11.

VIDEOGRAPHY

Chai, Jing 柴靜導. (2015). *Under the Dome*. (A documentary film in Chinese, with English subtitles). Youtube, https://www.youtube.com/watch?v=T6X2uwlQGQM&list=PLlRI77a4MFMYRqjbJ5UUSXourQpn3fD2v.

I

EXPLORING NON-ANTHROPOCENTRIC ONTOLOGIES AND NEGOTIATING SECULAR/ RELIGIOUS BOUNDARIES

Chapter 1

Rethinking Ontology with Equality of Life[1]

Jeffrey Nicolaisen[2]

I. BACKGROUND

The modern regime of human rights enjoys broad acceptance in global politics. However, as all forms of law and ethics, the origins and logic of human rights are contingent and historical. International organizations such as the United Nations Human Rights Council presume human rights to be an international norm and a universal set of values. The presumption of the universality of human rights is codified in the Universal Declaration of Human Rights (UDHR). While human rights are often presumed to be universally applicable, in actuality, the ontological assumptions that they represent derive from a Christian cosmology and do not necessarily represent the interests of all groups of people, especially those whose cosmology departs significantly from that of Christian monotheism. The philosopher with whom human equality may most be associated is John Locke (1632–1704). Locke argued that human equality derived from human rationality, a capacity that was provided by God in order to deduce the existence of God. While Anglophone culture has moved beyond expressing human rights in directly Christian terms, upon closer examination, the basis for the equality of humans cannot be assured without a theological basis. Alternative forms of equality, however, exist, and in this chapter, I introduce "equality of life"—the idea that all sentient beings, including both human and nonhuman animals, are equal—as expounded by the Taiwanese Buddhist nun Shih Chao-hwei (釋昭慧) (1957–present).

Chao-hwei is one of the most well-known Buddhist activists in Taiwan. Along with a group of her followers, activists, and business leaders, she founded the Life Conservationist Association (LCA; *Guanhuai Shengming Xiehui* 關懷生命協會) in 1993. At the core of LCA's mission is the promotion of equality of life through political lobbying and education. In addition to their activism, both LCA and Chao-hwei paved the way for academic research and debate on animal protection in Taiwan by putting traditional Buddhist thought on equality of life in conversation with international discussions on animal welfare and animal rights. In particular, one of LCA's earliest projects was the translation of Peter Singer's 1975 book *Animal Liberation* into Chinese. The publication of Xiangsen Meng and Yongxiang Qian's Chinese version *Dongwu Jiefang* in 1996 provided both academic and lay Chinese audiences access to the book widely considered to have laid the foundation for the modern animal protection movement. In 1994, shortly after cofounding LCA, Chao-hwei also began an academic career first at Fu Jen Catholic University and later at Hsuan Chuang University, where she is currently a professor of religious studies. Her scholarly work has focused on developing a modern applied Buddhist ethics through engaging with the style of philosophical discourse rooted in European traditions. In particular, she has maintained relationships with leading Anglophone animal welfare and animal rights theorists such as Peter Singer and Tom Regan and is currently working on a project to publish a dialogue on a broad range of ethical issues with Singer. As equality of life encompasses both human and nonhuman animals, Chao-hwei's scholarly work and activist endeavors extend beyond nonhuman animals and include topics that affect both human and nonhuman animals, such as gender equality, same-sex marriage, animal protection, gambling, and nuclear power. Chao-hwei has also been successful in forming international political alliances with organizations such as the World Society for the Protection of Animals (WSPA) and the International Network of Engaged Buddhists (INEB) in which she is one of four patrons, along with the Dalai Lama, Thich Nhat Hanh, and Maha Somchai Kusalacitto. Chao-hwei's combination of international political engagement and dialogue with European traditions makes equality of life a particularly relevant alternative form of equality to juxtapose with Lockean human equality.

Chao-hwei directly contrasts "equality of life" with human rights and advocates for animal rights and environmental rights as legal tools for advancing the principle of equality of life. She develops equality of life according to the fundamental Buddhist principle known as the "law of dependent arising" (*yuanqi fa* 緣起法)—the principle that all things originate according to causes and conditions—and upon this Buddhist basis for equality, neither the concept of rights nor the concept of nature can be established. However, in a legal system created with a logic based on the language of rights and nature, Chao-hwei still adapts to make the legal protections of rights useful where their interests align with appropriate legal standards.

To fully tease out the differences between human equality and equality of life, I take a deep look at the ontologies from which these notions of equality arise and the power dynamics that the international legal system propagates. In so doing, I challenge the applicability of the categories of nature and religion in discussing Chinese teachings by examining the Christian ontology that produced these categories. I demonstrate how the separation of church and state and the freedom of religion serve to suppress nontheistic teachings and propose applying the comparative methods of political ecology as a way to put various teachings on equal terms. Through analysis of Chao-hwei's exposition of equality of life, I show how an international politics based on Lockean rights forces non-Christian cultures to adapt legal concepts on a provisional basis for the protection of their interests and how new legal approaches such as the rights of nature bring new opportunities for alternative forms of equality.

II. THE CHRISTIAN ONTOLOGY OF HUMAN EQUALITY

In chapter 1 of Genesis, God created the world in seven days. He created heaven and earth on the first day, plants on the third day, animals of the sea and sky on the fifth day, and animals of the earth on the sixth day. While God made humankind along with the other animals of the earth on the sixth day, humankind held a special place in creation:

> So God created humankind in his image,
> in the image of God he created them;
> male and female he created them.
> God blessed them, and God said to them, "Be fruitful and multiply, and fill the earth and subdue it; and have dominion over the fish of the sea and over the birds of the air and over every living thing that moves upon the earth." (Genesis 1: 26–28; Coogan et al. 2010: 12–13).

The first chapter ends with God's day of rest on the seventh day.

In chapter 2 of Genesis, God animated both plants and humans, but in a notably different way. God created rain so the plants could sprout, "then the Lord God formed man from the dust of the ground, and breathed into his nostrils the breath of life; and the man became a *living being*" [my emphasis] (Genesis 2:7; Coogan et al. 2010: 13–14). In the New Revised Standard Version (1989) quoted here—the scholarly translation most often used by Anglophone scholars—the original Hebrew *nephesh hayyah* is translated as *living being*, but when God created the garden of Eden, the same term *nephesh hayyah* when used in reference to nonhuman animals is translated as *living creature* (Genesis 2:18–20; Coogan et al. 2010: 14). In the King James Version (1611), the difference in translation is even more stark: *nephesh hayyah* is translated as "living soul" for humans and "living creature" for nonhuman animals (Genesis 2:7 and 2:19; *Holy Bible* 1). The human being, here translated as "man," is the dust that becomes *nephesh hayyah* when animated by the breath of God. The Hebrew highlights the earthly material composition of the human being and does not divide the human into an earthly body and a divine soul (Coogan et al. 2010: 13, footnote 7). Rather *nephesh hayyah*, which means "animated creature," refers both to human and nonhuman animals, suggesting they both possess the breath of God (Clough 2012: Chapter 2). While theological scholars such as David Clough—from whom I am drawing much of this Biblical interpretation—are doing the constructive work to reclaim the status of the nonhuman animals in Christian discourse, the ontological distinctions that medieval and Enlightenment scholars constructed continue to have profound influence today.

According to Genesis, there are at least three types of beings. First was God, who is distinct from his creation. Second, there was all of God's creation, including both plants and animals. Third, there was "man" later joined by "woman." Ontologically, man is distinct from the rest of creation in at least two ways: in that he has dominion over the rest of creation and in that he is made in the image of God. The medieval and Enlightenment European scholars who constructed the intellectual foundation for the most popular forms of modern science and Christianity found the differences between "fish of the sea," "birds of the air," and "wild animals of the earth" or between plants and animals only as secondary to the threefold distinction. Convention further separates the human from the rest of creation by shortening "the rest of creation" to simply "creation."

In the Bible, the distinction between humans and God is stressed repeatedly, especially in the story of Jesus himself, in which God descends to inhabit the body of a man. Jesus' breach of the boundary between God and man is exactly what provides salvation. The otherwise impermeability of the boundary is a necessary condition for the exclusivity of the redemption that Christ offers. On the other hand, the distinction between man and animal was a subject of theological discussion in medieval and early modern Europe. While proofs of the distinction were various, the ontological distinction was not especially contested, and more often than not, simply presumed. Borrowing from Aristotle, the great Catholic scholar-saint Thomas Aquinas (1225–1274) distinguished humans from other creatures by rationality, which in Aquinas's Christian ontology took the form of a rational soul. For Aquinas, nonhuman animals were irrational "instruments" or "slaves" (7–12). Similarly, René Descartes (1596–1650) argued that animals are nothing more than machines, or "automata." Descartes succinctly concluded his observation with the description of two errors that implied the three-part ontology: "after the error of those who deny the existence of God, an error which I think I have already sufficiently refuted, there is none that is more powerful in leading feeble minds astray from the straight path of virtue than the supposition that the soul of the brutes is of the same nature with our own" (17).

For both Aquinas and Descartes, it was the "rational soul" that distinguished humans from animals. Animals, as with all of creation, were instruments to be exploited by the rational souls of humans.

In the second half of the seventeenth century, the capacity of rationality that, for European Christians, distinguished humans from animals became the theological basis for the new concept of human equality. At the time, human inequality, in which monarchs were by nature superior to their subjects, was the presumption in Europe. In that context, John Locke (1632–1704) provided a radical concept of human equality that formed the intellectual basis for human rights and modern liberal democracy. The legal scholar Jeremy Waldron identifies rationality as the criterion that in Locke's writing qualifies humans for equal rights. For Locke, the capacity of rationality by which God's creatures could deduce God's existence was the criterion for equality, and the capacity for this sort of rationality was specific to the human. In other words, the capacity to deduce the existence of God is the minimum criterion for equality, and any intelligence beyond that capacity does not enhance the status of any human above any other. The knowledge of God is sufficient to understand that there is a divine power who enforces a divine law with divine rewards and punishments. This knowledge then is sufficient to ensure that a citizen will behave virtuously.[3]

The problem with this equality is that there is a major exception that Waldron carefully demonstrates. Toleration does not apply to atheists. Rationality is the tool that God gave humans to deduce the existence of God, but those who have not yet utilized this rational capacity to realize there is a greater power do not yet qualify for full equality. Perhaps, in Locke, the true criterion for equality is belief in God, not rationality. In either case, however, the privilege of humans is given by God, because rationality itself was given by God so that humans could recognize him.[4]

Waldron argues that, unlike in Locke's era, human equality is now taken for granted, and people rarely reflect on the basis of human equality. Locke's argument for equality is the strongest available, but if we examine it closely, Locke's basis for equality is fundamentally a Christian one. Locke claims that "the atheist cannot really grasp the basis of the inalienability of human rights" (Waldron 2002: 227), so we may suspect that a non-Christian or nontheistic ontology would produce a different type of equality all together.

While Locke is the key founding theorist of human equality and human rights, there is more at stake than just political theory because the cosmology that emerges in the Enlightenment also forms the basis for science. Descartes's argument that animals were nothing more than automata was groundbreaking because it postulated a mechanistic universe that paved the way for scientific study. Descartes's contemporary Nicolas Fontaine (1625–1709) provides a description of some early studies that resulted from this Cartesian ontology:

> They administered beatings to dogs with perfect indifference, and made fun of those who pitied the creatures as if they had felt pain. They said that animals were clocks; that the cries they emitted when struck, were only the noise of a little spring which had been touched, but that the whole body was without feeling. They nailed poor animals up on boards by their four paws to vivisect them and see the circulation of the blood which was a great subject of conversation. (Rosenfield 1968: 54)

Descartes's rather radical view that nonhuman animals were incapable of feeling pain was not the only view, but the idea that there were irreconcilable differences between human and nonhuman animals was more common in Europe in Descartes's time. Locke did not reject the idea that nonhuman animals have feeling or even some limited ability to think. Instead, he maintained that they had no soul:

> If your lordship allows brutes to have sensation, it will follow either that God can and doth give to some parcels of matter a power of perception and thinking; or that all animals have immaterial, and consequently, according to your lordship, immortal souls as well as men; and to say that fleas and mites, &c. have immortal souls as well as men, will possibly be looked on as going a great way to serve an hypothesis. (1824: 76)

Clough argues that the anthropocentricism in early modern Europe related more to the development of technology, and the imperial project of exploring unknown lands and acquiring knowledge in order to exercise dominion over nature, rather than to rigorous Biblical interpretation (chapter 1). Indeed, Locke's interpretation that only humans possessed souls corresponds with the contemporaneous translation of Genesis in the King James Version but not the original Hebrew. The impulse for colonialism was a key motivation for Locke and his contemporaries, and the acquisition of knowledge through science was inescapably entangled with the colonial project.

Locke's justification of human equality enabled science by absolving scientists of ethical responsibility toward nonhuman subjects and even some human subjects, such as atheists. For example, the fundamental principle of conservation biology to protect biodiversity is generally linked with sustainable development for the purpose of human flourishing and delinked from the individual interests of nonhuman beings such as animals. For Clough, the failure to protect animals is a theological problem: "To fail to pause to attend to this part of God's creation would be to judge that, unlike human beings, animals have significance only as part of the ecosystems to which they belong, rather than being worthy of attention as individuals, communities and species" (Introduction, electronic edition). However, secular discourse now generally presumes human equality and human rights without examination of their Lockean Christian root. Lockean political theory enabled the mechanistic Cartesian cosmology to prevail for nonhumans, which continues to be reflected in present-day scientific practice. At stake in considering the jurisprudence is also the foundations of science, and it is in this context that I examine Chao-hwei's treatment of equality of life.

III. EQUALITY OF LIFE

Shih Chao-hwei recognizes the three-part classification of beings, or three-part ontology of Christianity, and proposes a Buddhist alternative. In her words, "Compared to Christianity's theory that God, humans, and animals have different statuses, the Buddhist theory of equality among sentient beings provides a strong theoretical foundation for treating humans and animals with equal kindness" (2014a: vi). Chao-hwei recognizes nonhuman animals as part of a more inclusive category she calls "life" (*shengming* 生命), or "sentient beings" (*zhongsheng* 眾生). The "equality of sentient beings" (*zhongsheng pingdeng* 眾生平等) is a common and well-known Buddhist concept in Taiwan, commonly called the "equality of life" (*shengming pingdeng* 生命平等). Chao-hwei is one of its most fervent advocates who both promotes it through her activism and develops it in her academic writing. The term *zhongsheng* refers to "all living things," but in the context of "equality of life" or "protecting life," "living things" or "life" frequently refers to "all sentient beings":

> The word *life* in the term *protecting life* still primarily indicates the sentient beings with the capacity of awareness, that is, animals. It is not at all the case that we do not need to protect and cherish plants and inanimate objects, but the capacity to be aware is not as strong as that of animals, and most have the capacity to grow back after being cut. Therefore, although the Buddhist precepts also require the monks and nuns to protect and cherish plants—not allowing them to be arbitrarily cut down—the precepts still primarily make animals with the capacity of awareness the object of ethical concern. (2019: 314).

"Life" in the aforementioned passage refers to "all animals with the capacity of awareness," and I adopt this definition of life when I refer to "equality of life." In Chao-hwei's writing and in the concept of equality of life in Taiwan, there is little distinction between "life" and "sentient beings."[5]

Chao-hwei's concept of equality is based on Buddhist principles, which she divides into three forms of equality:

1. Life uniformly has the capacity of awareness. We must equally respect the experience of suffering and joy and the strong aspiration to escape suffering and achieve happiness. This is the essential meaning of *equality of sentient beings* (*zhongsheng pingdeng*).
2. All life also arises, ceases, changes, and transitions according to causes and conditions. This is nothing other than the fact that, among distinctive elements, the many things produced by causal conditions have no self-nature and are equal and non-dual. This is the essential meaning of the *equality of the nature of all things* (*faxing pingdeng* 法性平等).
3. Not only do the ordinary people inherit suffering and joy and transmigrate through birth and death according to causes and conditions, sages also attain nirvana and complete the path to buddhahood according to causes and conditions. Under the law of dependent arising, the ordinary person has the possibility of liberation, and also has the possibility of becoming a buddha. This is the essential meaning of *equality of Buddha nature* (*foxing pingdeng* 佛性平等)[6] (2019: 320–321).

Each of Chao-hwei's three forms of equality contrasts with Locke's human equality in important ways. In regard to equality of sentient beings, or equality of life, as opposed to Lockean human equality's criterion of rationality, awareness is Chao-hwei's criterion for beings to qualify as sentient beings, or life. Thus, as an animal, a human, and any other sentient being including a God are all sentient, they do not require any distinction in status. Chao-hwei's classification system only includes two types of beings, sentient and nonsentient, and as we will see further, even the distinction between these two types of beings is ultimately only provisional.

According to Chao-hwei's second form of equality, equality of the nature of all things, no being can hold a special ontological status above others as God does in Christianity because all beings arise in the same way. Unlike the Christian ontology in Genesis, Chao-hwei flatly rejects any fundamental nature of beings, especially any form of original cause. In fact, Chao-hwei rejects any form of ontology whatsoever. To understand what she means by ontology, we must examine the genealogy of the Chinese gloss of ontology, *bentilun* (本體論). Taiwan's *Ministry of Education Chinese Dictionary* defines *bentilun* as:

> One branch of philosophy. Founded by Aristotle in the fourth century B.C. Primarily researching the commonality of all things, and the special characteristics that this commonality possesses. Also called "metaphysics."[7]

As may appropriately fit this Aristotelian definition, the Chinese character compound for *bentilun* means the "discourse on original essence." Aristotle's theory of causality traces back to an "Unmoved Mover." Drawing from Aristotle, Aquinas rearticulates the "Unmoved Mover" as the "Original Cause," which for Aquinas is God. Chao-hwei rejects the premise of searching for an "original essence" entirely:

> Among Aristotle's "first cause" or "first principle," Confucianism's "mandate of heaven," Daoism's "way" or "spontaneity," Moism's "will of heaven," the God of the Old and New Testament, Islam's Allah and Brahmanism's Prajāpati, or Mahā-Brahman, some possess personhood, while others are non-personal essences.[8] In this way, some come from tracing metaphysical origins through inference while others come from intuitive mystical experience . . . . Among these various religions and philosophies, Buddhism appears very special. It is not built on a metaphysical substance, and it did not establish the authority of heavenly revelation, but rather ascertained the law of the arising and ceasing of all things: dependent arising. In this way, it of course avoided some of the blame that some ontologists must face. For example, why is it that everything needs to arise due to a cause, but the "first cause" can exist by itself with no cause?[9] (2003: 47–49)

In other words, all things are equal exactly because they lack personhood or non-personal essences, or lack what in Buddhism is called "self-nature."

The third form of equality—equality of Buddha nature—introduces both a form of equality and a form of pragmatic hierarchy relevant to the obligation to protect life. A buddha is an "awakened one," or someone who has achieved nirvana. Nirvana refers to the extinguishing of self, like the blowing out of a candle. Buddha nature is the ability of all sentient beings to awaken to the realization that there is no self in order to liberate themselves from ignorance. "Since the term *life* in *protecting life* indicates sentient beings that have the capacity of awareness, even if animals other than humans are flying, squirming, or wriggling sentient bugs, each and every one fits into the scope of ethical concern, and it is not appropriate to favor one at the expense of another" (2019: 313). Sometimes, however, the interest of two or more sentient beings may come into conflict, and one must make a judgment about the weight of the relative interests of different forms of life:

> Originally, in principle, all-encompassing compassion extends to all sentient beings, but in a situation in which causes and conditions are limited, practical work must have an order of priority . . . . Buddhist disciples can draw up a Middle Way principle starting from the near and small and expanding outward, according to the differences in value between animals and humans and between common people and sages. (2019: 325)

In other words, all sentient beings are equal because they are capable of become buddhas, but the distance from buddhahood is not the same for all beings. Nonhuman animals are at a lower realm of existence in which they must be reborn as a human to practice the teachings of Buddhism. Humans are capable of achieving buddhahood, but sages are further along the path. Then, in pragmatic terms, those closer to buddhahood may deserve greater weight when the interests of two otherwise equal beings must be weighed against each other. Conversely, beings that are closer to buddhahood also possess greater capacities for moral reasoning and compassion. The greatest sages may "abandon reasonable self-defense in order to protect others," and the human sense of compassion obligates humans to protect nonhuman animals even if nonhuman animals do not have the capacity to return the favor (2019: 323).[10]

Chao-hwei's three equalities collapse the three-part classification consisting of animals, humans, and God in the Christian ontological model, but a two-part classification of sentient and nonsentient beings remain. Yet, consciousness itself is also subject to the law of dependent arising. "According to the theory of dependent arising, consciousness is not the most fundamental principle, and there is no way to prove that it comes from external divine revelation. Rather, it is the product of the reciprocal blending of subject and object" (2019: 319). Another way to think about dependent arising is "that every single sentient being is a body existing in the form of a network connected by infinite causes and conditions" (2019: 319). Thus, "all [sentient beings] are interdependently united with oneself, just like one body. At this stage, protecting life is not only simply a feeling, or a thought of benefitting others. It has become an enhancement of our consciousness" (2014a: 122).

Then Locke's concept of equality and Chao-hwei's concept of equality are incommensurable. Locke's concept of equality underlies the modern concept of human rights and remains the prototype for discourse on rights. Locke was one voice in a chorus of European intellectuals that promoted a natural theology and natural rights. The natural theology evolved into the modern classification of disciplines that separated the human (humanities and social sciences) from nature (natural sciences) and from God (theology). As the study of the natural world, transitioned from the study of God's creation to nature, the shift in terminology from "creation" to either "nature" or "environment" did not fundamentally change the three-part classification.[11] Humans were separate from nature, just as they were separate from creation. While Locke's separation of church and state, isolated theology to the private sphere, religion remained the realm of the "supernatural" or "superhuman"— that which was separate and above nature or humans.

In both Buddhist and classic Chinese cosmologies, there is no comparable concept to nature, as a realm of the mundane or the irrational nonhuman distinct from the ensouled or rational human. As a result, Master Chao-hwei borrows from Peter Singer's *Animal Liberation*, and explicitly rejects that rationality can serve as the criterion for a concept of equality:

Supposing that animals are a form of life, what reason do we have to equate these animals with plants and inanimate objects, only treat them as the background or the resources in an environment occupied by humans, and not conduct the principled ethical consideration described above? Supposing the criteria of the use of ethical principles are based on special characteristics (such as rationality) that humans do not share with other animals, then how do we also face the sharp skepticism of the animal liberationists? Can the treatment of people that have an IQ approximately equal to (or less than) an orangutan, dog, or cat—fetuses, infants, the learning disabled, and mentally handicapped—compare with that of animals? (2019: 317)

The 1948 UDHR declares, "All human beings are born free and equal in dignity and rights. They are endowed with reason and conscience and should act towards one another in a spirit of brotherhood."[12] Locke's criterion of reason is defined as an endowment of the human species. The endowment of conscience also reflects the virtue that Locke argued derived from that rational capacity to deduce the existence of God. Singer's argument that not all humans are more rational than all nonhuman animals challenges the presumption in the UDHR that all humans and only humans are equal, especially in regard to the criterion of rationality.

In her 2001 essay "Environmental Rights and Animal Rights,"[13] Chao-hwei challenges the foundations of rights and advocates for both environmental and animal rights. She begins by exploring natural rights, which in the standard Chinese gloss translates as *tianfu quanli* (天賦權利). As defined in the *Ministry of Education Chinese Dictionary*, *tianfu* means "God-given" (*shangtian fuyu* 上天賦與), so *tianfu quanli* translates directly as God-given rights.[14] In other words, the divine—and particularly Christian—provenance of natural rights is more explicit in the Chinese term than in the English term. This feature of Chinese is important because it plays out in the presumption that the rights-based system of thinking is fundamentally an artifact of Christianity. Within that context, Chao-hwei argues that environmental rights formed in the context of human exceptionalism. Humans realized the threat of environmental issues to human health and sustainability, and while their interests were in human rights, such as the right to health, they stated the new rights in the form of environmental rights. While she supports these protections, she also questions the conceptual feasibility of the attribution of rights to the environment since the environment does not possess the "rationality and conscience" that serve as the criterion for natural rights. She also asks whether environmental rights must take the form of "the establishment of a contract between God and humans"[15] (2014a: 316), and whether such a contract would be convincing to atheists. In answering these questions, Chao-hwei acknowledges the benevolent intentions of natural or God-given rights but also holds that there is no proof that they exist. The concept of

"God-given" (*tianfu*) is associated with Creationism and can only be affirmed through faith.[16] Significantly, Chao-hwei describes Buddhism as atheist, and declares that the Buddhist value of the "protection of life" (*husheng* 護生) is not based on the concept of rights as it is not given by God, but rather derives from dependent arising.[17]

Chao-hwei's concern is that the hopes of animal rights activists to achieve legal personhood for animals will be thwarted by anthropocentric theories of ethics. If they believe anthropocentric philosophers, they will fall into a trap of thinking that animals do not deserve protection because they are not capable of rationality. Despite her reference to Singer's argument that humans are not uniformly more intelligent than nonhuman animals, Chao-hwei does not deny that humans *on average* have greater abilities for rationality and morality than other animals. Instead, she argues that the mistake of philosophers such as Immanuel Kant is that they made the capacity of reason, conscience, or moral reasoning the criterion for ethical consideration. She believes that Christianity-influenced philosophers conflated the capacity for agency with the criterion for ethical consideration. For Chao-hwei, the capacity of sentient beings to suffer is the correct criterion for moral consideration, but the obligation to protect sentient beings falls on humans, because humans are capable of moral reasoning. Nonhuman animals may cause other beings to suffer, but because they are not capable of moral reasoning, we cannot expect them to behave morally. In this regard, the Buddhist concept of the protection of life is quite different than human rights theory, and other European moral system such as Kant's deontology. Chao-hwei does not believe animal rights advocates should be discouraged from implementing the protection of life, and while she distinguishes natural rights from the protection of life, she still supports legal rights for animals, such as legal personhood granted to nonhuman animals.

For Chao-hwei, animal rights are distinct from environmental rights. Environmental rights stem from a broader concept of dependent arising and no self. The "environment" is the set of causes and conditions that support life. In this sense, the protection of life demands the protection of the environment. Recognizing that all life depends on a multitude of causes and conditions requires that biodiversity be respected in order to support sentient life. While she believes environmental protections, such as those represented by the Convention on Biological Diversity, in large part, stemmed from anthropocentrism, she still supports the legal protections they provide. However, because environmental rights only protect the causes and conditions that support life, she argues that it should be second in status to animal rights, which directly protects sentient beings.

To fully understand Chao-hwei's position, it is important to draw a distinction between two types of rights that she places under the category of environmental rights. In the Anglophone legal literature, the term *environmental rights* has come to be associated with the "human right to a healthy environment."[18] Chao-hwei notes that the anthropocentric movement for the human right to a healthy environment began in German around 1960. While she supports this form of human rights because they still provide some protection to the environment, she aims to provide a Buddhist basis for the value of the environment independent of humans. She argues that the connotation of the term *environmental rights* changed from the anthropocentrism of sustainable development to the ecocentrism of radical environmental advocates in the 1970s. As she specifically mentions an early concern for plants, she appears to be referring to Christopher Stone's 1972 essay "Do Trees Have Standing—Toward Legal Rights for Natural Objects." In the essay, Stone provocatively asserts that nonhumans such as trees may have legal standing independent of their human use. In the Anglophone literature, the form of rights that recognizes the value of nonhumans independent of human use has come to be known as the "rights of nature," a concept that only gained currency after the publication of the Catholic theologian Thomas Berry's 1999 book *The Great Work*.[19] While in her 2001 essay, Chao-hwei admits environmental rights had not been very successful at the constitutional level, the rights of nature movement has achieved several victories since the publication of her essay: Ecuador's 2008 constitution and Bolivia's 2009 constitution both recognize the rights of nature. The ratification of Ecuador's constitution in particular marked a major landmark in the international rights of nature movement and served as a bellwether for a wave of legal changes at various levels of government from treaty negotiations in New Zealand to court decisions in India.[20] Ironically, the "rights of nature" translates directly into Chinese as the exact same term used for the God-given human rights known as *natural rights* (*ziran quanli* 自然權利). Conversely, Chao-hwei's term *huanjing quanli* (環境權利) can translate both as environmental rights (the human right to a healthy environment) and the rights of the environment (the rights of nature). Despite the dual meaning of the term, Chao-hwei clearly distinguishes between the two types of *huanjing quanli*, and builds a Buddhist case for the subjectivity and non-use value of the environment corresponding with the rights of nature's core concept of granting personhood to nonhuman entities.

Chao-hwei only accepts animal rights and environmental rights in a provisional sense to support legal protections for animals and the environment. The translation problems are not simply a matter of linguistic inconvenience. Chinese cosmologies including those of Daoism, Buddhism, and Confucianism lacked an all-powerful God that created all of creation and existed outside of it. While Chao-hwei is very explicit about reformulating the concepts of rights, she is less explicit about reformulating the concept of environment.

First, she took the nonhuman animals out of the concept of the environment completely, making the boundary of environment expand to encompass all sentient beings. Second, she claimed the subjectivity of all nonsentient things as they are the causes and conditions for all sentient beings, joining with them as a single body. The environment, then, is the collective karma of all sentient beings. These much deeper ontological differences merit brief consideration.

IV. THE ONTOLOGY OF NATURE

In China, the concept of *huanjing*, a term that has become synonymous with the English word *environment*, represents an ontological divide that did not exist in early modern China. The compound term *huanjing* (環境) combines the character for "ring" or "circle" (*huan* 環) with the character for "terri-tory" (*jing* 境). In classical Chinese, it functioned as a noun which meant "the outer limits of a territory," or as a verb "to enclose the outer boundaries of a domain." Historically, the term was used in the context of identifying who exercised dominion over an area, and strongly implied the active colonization of that domain. For example, the term appears in the "Yuan History, Biogra-phy of Yu Que" (*Yuanshi*, Yuque zhuan 元史 · 余闕傳):

> After arriving and occupying [the territory] for ten days, bandits arrived, and [Yu Que's forces] stood their ground and repelled them. Thus, [Yu Que] assembled the bureaucratic and military officials and discussed their strategy for fortification of the fields and military defense. [They] built berms and fences to *surround the territory* [huanjing, my emphasis], chose shiny armor and external shields, and ploughed and planted the interior.[21]

The term did not represent a divide between humans and nature, but rather a divide between a settlement and the threat of that which was outside, a threat serious enough to require a barrier. By the first half of the twentieth century, the new imported definition in which *huanjing* describes surrounding natural conditions established itself in notable works such as the foreword to *Mr. Lu Xun's Complete Compilation* (1938; *Luxun xiansheng quanji* 魯迅先生全集) by Cai Yuan-pei (蔡元培):

> Walking on the path of Shan-yin, the thousand rock-cliffs compete to be the most beautiful, the ten thousand gullies vie to flow, making people respond to their demands with no rest. There is this type of *environment* [huanjing, my emphasis].[22]

According to this usage of *huanjing*, the division that the term represents is no longer military barriers between the settlement inside and the threat outside, but rather, it becomes the divide between the natural and the human. *Huanjing* in this regard is both awe-inspiring and threatening—something separate from the human, but something on which humans must act to contain. In other words, in the first half of the twentieth century, Chinese culture began to adopt the ontological distinction between humans and the rest of God's creation. While the Christian distinction between humans and the rest of creation was recoded in terms of humans and the environment or humans and nature, the idea that there was a distinction between the two crept into the modern Chinese language and shared imagination.[23]

The neologisms that formed to translate environment's close twin "nature" revealed not only that the human-nature distinction was new to China but also that the human-God distinction, or mundane-sacred distinction, was also foreign to Chinese culture. The influential sixteenth-century Jesuit missionary Matteo Ricci interpreted the term *tian* (天) to mean a monotheistic God. However, as the anthropologist Robert Weller points out, until the past several decades, English translations of Chinese texts often rendered the term *tian* as "nature." Now the term is generally translated as "heaven" (Weller 2006: 21). Nonetheless, in compound form, it still refers to both God and nature. As mentioned earlier, the term for God in the UDHR was *Shang-tian*, while in the contemporary vernacular, the term *tianran* (天然) means natural, as in the products at a natural foods store. Similarly, the currently most common gloss of the word nature in Mandarin Chinese is *ziran* (自然). This term originally meant "spontaneous," "free," or "unencumbered" and functioned as one of the core concepts of Daoist teaching. For example, according to the *Laozi* (老子; sixth or fifth century B.C.), "Humans are ruled by the earth. The earth is ruled by heaven. Heaven is ruled by the Way. The Way is ruled by *ziran* [my emphasis]."[24] For Laozi, humans—along with the earth and heavens—were ruled by the spontaneous becoming called "*ziran*." The Japanese coined the new usage of *ziran* (Japanese, *shizen*) to translate the English term *nature* in the late nineteenth century (Shogakukan 2014). Then China imported the new usage of *ziran* along with the new term for religion (Chinese, *zongjiao*; Japanese, *shūkyō*) from Japan. In other words, it imported the human-nature distinction and natural-supernatural distinction simultaneously as a response to increased contact with European powers.

In his monograph on nature in China, Weller asks whether there is "an equivalent to the English word 'nature' in Chinese before the twentieth century," and concludes, the "monosyllabic answer is no" (Weller 2006: 21). In classic texts, Daoists, Confucians, and their contemporaries considered humans to be constituted by a single "hun" (魂) and a single "po" (魄), both of which were composed of the fundamental energy called "qi" (氣). According to a Confucian text called the "Meaning of Sacrifice" (*Jiyi* 祭義), "In the conversion of life [after death], the hun is called shen 神, and the po is called gui 鬼." "Shen" is generally translated as "god," and "gui" as "ghost," but these Anglicizing terms do not adequately represent the meaning in Chinese. Everyone possessed an energetic formation called a *shen* [i.e. *hun*] that survived them after death.[25] Currently, the most common hun-po theory is that each human has three hun and seven po. Nonhuman animals are similar but lack one of the hun.[26] One of the highest-ranked deities in the imperial register of sacrifices (sidian 祀典) Guan-gong was originally a human that became a deity after he died. Wenchang-gong (文昌宮), a deity of similar rank, was originally a serpent.[27] Chinese deities are most often the continued life presence of deceased human or nonhuman animals, quite different from a monotheistic God. Even the Greek and Roman gods were different as they were embodied, immortal, powerful beings, not the energetic residual of previously embodied humans or animals, although shen could also be quite powerful. Rather than the ontological split between God, humans, and nature, early Han cosmology understood all things to be composed of the same basic form of energy.

The relatively similar ontology of humans, animals, and *shen* was reflected in their relationships. In his study of the early Chinese textual treatment of animals through early Imperial China, the anthropologist Roel Sterckx demonstrates that nonhuman animals could be held up as paragons of virtue and be treated as subjects of the state as well as demonized as immoral brutes:

> The classic Chinese perception of the world did not insist on clear categorical or ontological boundaries between animals, human beings, and other creatures such as ghosts and spirits. The demarcation of the human and animal realms was not perceived to be permanent or constant, and the fixity of the species was not self-evident or desirable. Instead animals were viewed as part of an organic whole in which the mutual relationships among the species were characterized as contingent, continuous, and interdependent . . . . Through linking the comprehension of animals with the ruling of human society, the early Chinese presented the animal world as a normative model for the establishment of sociopolitical authority and the ideal of sage rulership. (Sterckx 2002: 5)

Not only did early Han civilization lack the distinction between animals, humans, and gods, but they also used the nonhuman animal behavior as a model for the structure of human society. As Mencius said, "The differences between humans and the birds and beasts are very small."[28] In fact, he exhorted his followers to "love parents affectionately, and care for the people. Care for the people, and be fond of animals."[29] In the Confucian concentric theory, animals were valuable and deserving of compassion, but filiality to parents and concern for humans could easily outweigh the interests of animals. Zhuangzi outlined a more egalitarian ideal for the relationship between humans and nonhuman animals:

> It is called the emancipation of the heavens (*tianfang*) . . . . The ten thousand beings live species by species, one group settled close to another. Birds (*qin*) and wild animals (*shou*) form their flocks and herds, grass and trees grow to full-est height . . . . In this age of perfect inner power people live the same as birds and wild animals, group themselves side by side with the ten thousand beings. Who then knows anything about "superior" or "inferior person"? Abiding in nonknowing (*wuzhi*), their inner power does not depart from them. Abiding in desirelessness (*wuyu*) is called unadorned simplicity (*supu*). Through unadorned simplicity people realize their innate nature.[30]

Zhuangzi's nonknowing, desireless abiding was to live without effort as an animal. Also notable in this passage is that the phrase "emancipation of heaven" could as easily be translated as "emancipation of nature," again demonstrating the lack of distinction between these categories. The role of humans was neither to dominate nature nor nonhuman animals, rather it frequently was to learn from animals and emancipate them.

On the other hand, a lack of ontological distinction between human and nonhuman animals could also lead to an equality of violence. Exotic plants, animals, and objects at the outer boundaries of homes or communities possessed special power, and humans sacrificed and consumed animals and plants to benefit from their power.[31] Han settlers in Taiwan were beset by attacks from both headhunting indigenous people and wild boar. While the indigenous practice of headhunting is much maligned by Han people, Han defenders of indigenous people have pointed me to cases of Han consumption of indigenous peoples in Taiwan. In some cases, even the meat of indigenous people was for sale at Taiwan's markets in the nineteenth and early twentieth centuries and eaten for its special power, just as the meat of exotic animals. In the second half of the nineteenth century, the physician George Mackay reported observing Han people consuming the brains of an indigenous man to enhance their intellect and bravery and creating a bone jelly from another indigenous man to treat malaria.[32] The will of *shen* could also be bent toward human advantage through proper veneration and offerings. In present-day Taiwan, the most well-known ritual animal sacrifice is the annual Shen Pig Ritual (*shenzhu jidian* 神豬祭典).[33] Contrary to Buddhist prohibition on killing sentient beings, the ritual is held at several temples that venerate Qingshui Zushi (清水祖師), a Buddhist monk who performed various great deeds.[34] The ritual involves sacrificing fattened pigs called "shen pigs" to Qingshui Zushi on his birthday, the sixth day of the lunar new year. The local people believe that Qingshui Zushi blesses the pigs so that the pigs may eat especially well, and they hold a competition in which the participant who raises the fattest pig wins a medal. In 2018, the winning pig was 892 kg, seven to eight times the weight of a normal pig. The ceremony demonstrates how the idea that *shen* need to eat just as humans do is passed down from early hun-po theories, and that sacrifice—including the sacrifice of animals—is a method of veneration and accruing favor with *shen*. The particular irony of the Shen Pig Ritual, however, is that as a Buddhist monk, we can expect would have been vegetarian and opposed to killing animals. This counterintuitive entanglement of traditions demonstrates how inextricable the multiple traditions are from each other, and how it is difficult to make generalizations based on a particular category such as Buddhism or Daoism. Qingshui Zushi is not only both a *shen* and a bodhisattva (a Buddhist practitioner that forgoes Enlightenment to save all beings), but also a tutelary *shen* of the Quanzhou region of Fujian Province. Both the consumption of indigenous people and the ritual sacrifice of *shen* pigs represent responses of Han settlers to threats to their lives. In the case of swine sacrifice, the ceremony performs the victory of the Quanzhou settlers over the native wild boar through the traditional rituals associated with the Han cosmology.

Chao-hwei, the Life Conservationist Association (LCA), and their allies have actively documented, publicized, and protested the offering of *shen* pig and similar animal sacrifice rituals. In one notable example, Chen Shui-bian, the president of Taiwan from 2000 to 2008, pledged to sacrifice a *shen* pig to the Hakka *shen* Yimin Ye (義民爺) if Yimin Ye would help him win the 2000 election and bring peace and prosperity to Taiwan. After his election victory, several farmers raised *shen* pigs for the president to sacrifice and called them "president pigs." Chao-hwei led a movement to stop the ceremony, arguing that violent rituals could not bring peace to the country but rather only peaceful means such as a pledge of vegetarianism could bring peace to the country.[35] However, despite a vocal opposition, Chen participated in the president pig sacrifice in 2003 to respect the ritual of the Hakka people (Pingguo ribao 2003).

Thus, the culture of blood sacrifice in Taiwan and China is strong to the present-day. People consume exotic animals to accrue their power and sacrifice other animals to curry favor of the *shen*. Traditional Chinese cosmologies offer both idyllic visions of human harmony with the cosmos and methods of manipulation of energetic forces for human benefit. These cosmologies are actually intimately entangled and inseparable from Chinese Buddhism, but as we can see, even if we hold them apart, they do not reproduce the three-part Christian cosmology creation-human-Creator or its secular equivalent natural-human-supernatural.

Buddhism introduced new ideas that added to the array of cosmological possibilities in Chinese culture, but it similarly lacked the three-part cosmology brought by Europeans. The Buddhist studies scholar Ian

Harris considered a variety of Indic terms as candidates for the term *nature*, including *saṃsāra, prakṛti, svabhāva, pratītya-samutpāda, dharmadhātu, dharmatā,* and *dhammajāti,* but found them all inadequate.[36] First, Harris noted that the distinction between nature and supernature arose in medieval Europe in the work of Thomas Aquinas and his contemporaries. These theologians believed that what could be observed in nature was not sufficient to explain the totality of the universe, and therefore they induced the existence of the supernatural. Second, Harris considered *saṃsāra* as a candidate for the term *nature. Saṃsāra* denotes a universe characterized by the continuous cycle of transmigration of all beings. All beings cycle between six categories of rebirth: hell beings, hungry ghosts, animals, humans, asuras, and devas. While *deva* may be translated as God and may share etymological roots with the word divine, it does not represent the same concept as a Christian God. Devas die and are reborn in the cycle of transmigration. They have at one time in the past been animals and at another time humans and will be animals and humans again in the future, just as the beings in all categories of rebirth continuously change between these categories. This cosmology in which all sentient beings are in the same state of samsara negates the possibility of a nature-supernature distinction. Harris concludes that the other Indic terms also fail a similar test of their equivalency to the concept of nature.

After Buddhism arrived in China around the first century AD, it became extremely influential, as many Chinese people accepted the idea of transmigration, its accompanying concept of karma, and the teachings of compassion for all sentient beings. Karmic theory asserted that one's actions not only affected one's current and future lives but also the direction of the collective cosmos. Karma shared some affinities with the Chinese concept of sympathetic resonance (*ganying* 感應). Sympathetic resonance involved a universe that responded to human behavior, as represented by expressions such as "*Tian* and humans are joined together as one" (*tianren heyi* 天人合一). In both Buddhism and native Chinese traditions including Confucianism and Daoism, the universe was in a reciprocal relationship with humans, and behavior and intention cocreated the universe. There was no divine being interceding from outside of the system. The concept of transmigration was new for Chinese people when Buddhism arrived in China, and the teachings of compassion for all sentient beings offered a new cosmological emphasis on concern for nonhumans. I have found that in my own interactions with Han people in present-day Taiwan, transmigration is the common default way of apprehending life and death, although it is entangled with the traditions developed from hun-po theories. The concept of the universe resonating with or responding to moral behavior is also one of the most foundational notions. One college student at a "religion and social movements" training camp I attended summarized the shared cosmology as "Evil has evil consequences, good has good consequences" (*e you e bao, shan you shan bao* 惡有惡報，善有善報). Other participants at the retreat agreed with this formulation. This shared cosmology also explains why—as the Religious Experience Survey in Taiwan (REST) demonstrates—those with higher levels of education and more formal exposure to Confucian teachings may begin not only to affirm experiences with Confucian concepts but also Buddhist concepts such as karma (Voas 2013).

This recompensatory system included concern for nonhuman animals. From the Han period, as traditions of the Daoist masters developed, Daoists opposed the profligacy of eating meat and ritual sacrifice. In the period from the first century to the fifth century CE, the Daoists codified these practices, and the government began to see abstention from meat rather than animal sacrifice as a way to produce positive cosmic resonances. Some Confucians were also committed to care for animals, including the sixth-century Emperor Wu of Liang (Liang wu di 梁武帝, 464–549 BC; r. 502–549 BC), who committed himself to vegetarianism as part of Confucian practices of mourning. The Buddhist sutra called the *Brahma Net Sutra* (*Fanwang jing* 梵網經, app. fifth century BC) appeared in China at this time, and included a precept prohibiting eating meat. Emperor Wu himself took the precepts and forbade Buddhist monks and nuns from eating meat. Thus, vegetarianism solidified as a way of gaining cosmic or karmic merit through multiple Chinese traditions, even though it came to be especially associated with Buddhism. This early history produced a foundation for Buddhist vegetarianism as well as the Vegetarian Jiao (*Zhaijiao* 齋教) that became popular in Taiwan during the Qing Dynasty. The practice of vegetarianism remains a common method of gaining merit in Taiwan. For example, in my own experience, at a retirement event for a Legislative Yuan representative with a history of sponsoring legislation for people with disabilities, I met a man on the board of an autism association. He became vegetarian as a way to help his son who had a severe form of autism. He was not Buddhist but saw vegetarianism as a way of reaping cosmic merits that he could transfer to his son. This form of practicing vegetarianism is exactly what Chao-hwei proposed to Chen Shui-bian as an alternative to sacrificing the president's pig. While the prescription of vegetarianism was a Chinese innovation, Buddhist theories of transmigration and karma provided the strongest foundation for compassion toward nonhuman animals, as illustrated by practices such as the prohibition on killing and the "release of life" ceremony.[37]

Thus, Chinese cosmologies including those of Daoism, Buddhism, and Confucianism lacked an all-powerful God that created and existed outside of creation. The *shen* and bodhisattvas were an integral part of a cosmos that responded to moral behavior and ritual, and they themselves generally had been nonhuman animals or humans in previous lives. Humans equally could not be held apart from the system as the rulers over a nonhuman nature, because they were in a reciprocal relationship integrating humans with tian. Nonhuman animals were part of the same energetic or cosmic universe and worthy of moral consideration. Thus, there was no native concept of "environment" or "nature" that correspond with a nature-human-supernature distinction. Implicitly, then, when Chao-hwei uses the term *environment*, she is adapting a North Atlantic concept to a Buddhist cosmology, just as she is doing with the term *rights*. Chao-hwei clearly explains the theological provenance of natural rights, and rejects the idea that God granted any rights. She explicitly asks whether the concept of rights can be salvaged without the theological basis, and defends two new forms of rights—animal rights and environmental rights—based on the Buddhist principle of "equality of life." While she is very explicit about reformulating the concepts of rights, she is less explicit about reformulating the concept of environment. Considering that I have already shown that the concept of the natural environment does not draw on a premodern Chinese cosmology, we can also identify how Chao-hwei reformulates the modern notion of environment. First, Chao-hwei took the animals out of the concept of the environment completely, making the boundary of *"huanjing"* expand to encompass all sentient beings. Second, she claimed the subjectivity of all nonsentient beings as they are the causes and conditions for all sentient beings, joining with them as a single body. The environment, then, is the collective karma of all sentient beings.

V. THE ONTOLOGY OF RELIGION

Considering the failure of the tri-part nature-human-supernature ontology to describe Chinese cosmologies, there is one more significant issue to consider. As J. Z. Smith argues, in the transition from the theological study of religion to the anthropological study of religions, definitions of religion evolved from a theological definition based on a single supreme deity to an anthropological definition distinguished by the "supernatural" or "superhuman." He cites Melford Spiro's 1966 definition as representative of the anthropological definition: "an institution consisting of culturally patterned interaction with culturally postulated superhuman beings," with the further qualification that "religion can be differentiated from other culturally constituted institutions by virtue only of its reference to superhuman beings."[38] Whether the terms "superhuman" or "supernatural" are employed, they both rely on the hierarchical construction of the three-part nature-human-supernatural ontology, where humans dominate and are above nature and God dominates and is above both humans and nature. Thus, "superhuman" (above humans) and "supernatural" (above nature) both refer to the position of God in the nature-human-God cosmology.

Furthermore, Aaron Gross demonstrates that secular religious studies theorists confine even the realization of religion to humans. He examines Emile Durkheim, Ernst Cassirer, Mircea Eliade, and Jonathon Z. Smith, and identifies how they equate the emergence of rational or symbolic thought with the emergence of religion. In other words, just as in Locke and his contemporaries, rationality is what distinguishes humans from nonhuman animals. Gross writes, "One could even argue that Smith and those who advanced this direction of theory before him have failed to shift religion from the sphere of the divine sciences. After all, it is exceedingly difficult—and I tend to think impossible—to argue for an ontological distinction between humans and all other life that would justify limiting the study of religion to the sphere of the human without an appeal that extends beyond the domain of reason" (Gross 2014: 85). Thus, religion is about the supernatural, but the capacity to create it is what distinguishes humans from nature.

These arguments fail when there is no nature-human-supernature ontology, as in the case of all Chinese teachings. The Christian cosmology must be superscribed on non-Christian teachings in order to produce religion. In other words, cosmologies that do not include the same three-part ontology cannot be reliably classified as secular or religious because they do not include the relevant distinctions.

Smith is quite open about the flexibility of the definition of religion. Considering James H. Leuba's accounting of more than fifty definitions of religion, Smith concludes that all are equally valid: " 'Religion' is not a native term; it is a term created by scholars for their intellectual purposes and therefore is theirs to define." This statement may be correct, but it does not fully address the political consequences involved in defining religion. The definition of religion has real political significance, determining, for example, what organizations get tax benefits, whether an organization is subject to strict or loose accounting regulations, and what teachings may be a part of the compulsory curriculum. How scholars define religion is directly implicated in the resolution of these questions. Taiwan's Ministry of Interior (MOI) hired National Chengchi University's Graduate Institute of Religious Studies to consult them on how to define religion. According to Kuo Cheng-tian, a political scientist and member of the faculty who consulted the MOI, the Graduate Institute evoked the philosophy of John Locke in recommending a policy not to define religion.[39] Having reviewed Locke's theory of human equality, we can see that not only is the concept of religion based on a Christian cosmology, but the institution of human rights that protects the freedom of religion is also built on a Christian cosmology. Furthermore, as we recognize that Locke created religious freedom only for monotheists and that, as

Waldron argues, the only defensible basis for human equality lies in Lockean Christian theology, then we also must recognize that the institution of human rights is an ontologically Christian institution. The political implications of this problem are probably the major issue with which the field of religious studies must grapple.

For example, in Taiwan, a small number of educators argue that Falun Gong should not be taught in schools because there is no objective standard to confirm the content of the teachings (Chen and Huang 2004). However, if we apply that same standard to the curriculum on human rights and human equality, we would not only have to conclude that there is no objective standard to confirm these teachings but also that they are also a religion and should be prohibited from the schools. In fact, we have already seen that they rely on a Christian ontology, so this conclusion is stronger than the conclusion about Falun Gong. Natural rights are based on the endowment of the Creator, which fits the original definition of supernatural and superhuman, so the basis of these rights fits both Smith's theological definition of religion and his anthropological definition. Falun Gong, on the other hand, developed as a form of *qigong*, so it relies primarily on Chinese cosmology that lacks the nature-supernature distinction. It does not fit either the theological or the anthropological definition of religion. However, while organizations such as Falun Gong and the Buddhist organization Tzu Chi compete for access to Taiwan's Life Education curriculum through the indirect means of continuing education courses for teachers, human rights are a required part of the curriculum, just as Confucianism.[40] Human rights are also enshrined in Taiwan's constitution, so no legal argument can be made to eliminate human rights from the curriculum. Furthermore, this discussion of the educational curriculum is much more than a peripheral topic in regard to the interests of teachings, or "jiao" (教), because the function of teachings is to be taught. The privilege to teach is the privilege to survive as a teaching. International politics and the legacy of colonialism enable human rights to enjoy this privilege in Taiwan's mandatory curriculum, while Chinese teachings like Buddhism and Daoism do not.

Even Chao-hwei believes that human equality and human rights were a step in the right direction, but she argues for an expansion of the concept of equality to include all sentient beings. To be able to fairly consider an alternative form of equality as represented by the equality of life, however, we need to put it on equal terms with the existing form of equality. Since human equality is already enshrined in constitutional law and required as a part of mandatory curriculum across the world, it is unquestionably the hegemonic form of equality. The best we can do to maintain symmetry with the hegemonic form of equality is to make existing power relationships transparent. By demonstrating the Christian roots of human equality and exposing the power advantages it enjoys is not to denigrate its value, but rather a necessary step to compare it with an alternative that does not enjoy the same advantages—in other words, to put them in symmetrical relationship to the extent possible.

Research on alternative ontologies has been a trend in the field of political ecology in the past ten to fifteen years, with roots that go back even farther,[41] but there may be some reasons that this approach has been much more limited in the study of religion. The most productive area of scholarship has been the Amazonian basin. Anthropologists such as Phillippe Descola, Eduardo Viveiros de Castro, and Arturo Escobar have pioneered this new approach in that region, with later contribution from scholars such as Eduardo Kohn and Marissa de la Cadena. Notable scholars in other geographic areas such as Elizabeth Povinelli and Deborah Bird Rose in Australia have also made significant contributions. My point is not to list all the significant scholars in this field or review their scholarship, but rather to simply note that these scholars work on the ethnography of indigenous communities that cluster in certain geographical areas. As Smith noted, a world religion is "a tradition that has achieved sufficient power and numbers to enter our history to form it, interact with it, or thwart it," while all " 'primitives,' by way of contrast, may be lumped together, as may the 'minor religions,' because they do not confront our history in any direct fashion. From the point of view of power, they are invisible" (Smith 1998: 280). This invisibility may have been exactly what allowed anthropologists to take indigenous ontologies seriously, while the power of world religions may threaten existing power structures. According to the logic of the freedom of religion, taking religion seriously is theology (a misnomer for Chinese teachings) and thus not allowed in the public sphere. In the case of China, many minor traditions were destroyed as "superstitions" in the first half of the twentieth century, so major teachings such as Buddhism and Daoism had every reason to convince the government they were religions as a matter of survival. That status, however, locked them out of the domain of power and the public sphere, and tacitly prohibited what religion theory considered "theological" scholarship in the public domain.

In the same period that ontology scholarship has grown in popularity among in the field of indigenous studies, the "religion school" has grown in influence in Chinese religions. In an interview with Kuo Cheng-tian, he used the term *religion school* to refer to the emerging group of scholars that recognize the lack of a mundane-sacred (natural-supernatural) binary in Chinese teaching. They still retain the category of religion and the restrictions on scholarship that corresponds with the category. Some have even reasserted the category to show that Chinese traditions are not merely secular.[42] These approaches have shed light on the problems with categories of religion and the secular, but have not moved so far as to reject these categories themselves. Because the assignment of Buddhism and Daoism to the category of religion has been so broadly accepted, especially in the West, I argue for the category of "second religions" to recognize this broad acceptance while simultaneously holding them apart as ultimately to recognize that they can neither truly fit in the category of religion nor the category of the secular. In many ways the "religion school" has produced the basis for a transition to a new approach.

According to Robert Orsi, the Protestant reformer Huldrych Zwingli started a process that would ultimately lead to "God's absence from the world" when he claimed that Christ was only in the Eucharist symbolically, not literally. The Protestant "divide between presence and absence, the literal and the metaphorical, the real and the symbolic, the natural and supernatural, defines the modern temperament" (Orsi 2016: 37). Rather than starting the process of disenchantment with Zwingli, it could be traced to Catholics themselves when Aquinas and his contemporaries created the nature-supernature divide. Since Catholics created the category of the supernatural, Catholics at least acknowledged the category when modern scholarship under Protestant influence rejected the actual presence of God in the mundane world. On the other hand, recognizing the "presence of God" in Chinese teachings was completely nonsensical, as God in the Christian sense did not even exist in these teachings. As this natural-supernatural divide did not exist, colonizers and their local collaborators could define what was natural and supernatural by superscribing these categories on local teachings. By assigning Buddhism, Daoism, Confucianism, and other teachings to the category of religion, they made a judgment that these teachings were supernatural, rejecting their legitimacy in contributing to the understanding of nature. Where any truth value broke through to the secular, it needed to be authorized by mediators of truth who would appropriately "naturalize" the knowledge with secular ritual, as in the experiments involved in the scientific study of mindfulness and meditation. Even then, reference to the original teachings with terms such as "Buddhist" would need to be erased to allow the knowledge to enter the public sphere. Yet, the political conditions that mediated these processes such as human equality, human rights, the separation of church and state, and the freedom of religion belonged to the Lockean Christian ontology of the European Enlightenment.

The use of an approach that suspends the category of religion in order to take Chinese teachings on their own terms is just beginning. One pioneering work is James Miller's 2017 book *China's Green Religion*. Miller recognizes the three-part ontology I discuss in this chapter: "in this modern imagination, the realms of the supernatural, the natural, and the human are fundamentally distinct from one another, and the three disciplines of religion, science, and philosophy focus on each of these three realms respectively" (12). From that starting point, he presents Daoism as an alternative way of apprehending the pressing issue of sustainability, and argues that a Daoist cosmology offers a better way of approaching sustainability:

> The Daoist approach is better in three respects: first, as a basic paradigm for apprehending the world we live in, it fits better with the findings of evolutionary science, ecological science, and environmental science. Second, as a basic paradigm for orienting human life toward the world we live in, it provides a spirituality and a worldview that is creative and life sustaining. Third, as a mode of practical engagement with our world, it is profoundly relevant for the global quest to create an ethical framework that produces a flourishing world for the betterment of human life and the sustainability of the planet humans depend on for their survival. (18)

In relation to the goal of sustainability, Miller clearly believes that Daoism is better not only politically but also scientifically. In particular, Locke and his contemporaries formulated a concept of the human individual that was set apart and above nature. This model did not allow for the human to be fully engaged in a reciprocal relationship with an active and subjective nature, but rather situated the human in dominion over a passive and objective nature. In this doctrine, the human individual was not only "buffered" from the environment but also from other human individuals. Human rights doctrine is just one example of this buffering. As unalienable rights were endowed by the Creator, the rights of the individual did not depend on relationships with other humans or nonhumans in the world, and thus rights conceptually buffered the individual from social relationships as well as the environment. Charles Taylor calls this model of the human the "buffered self." As opposed to the Enlightenment model of an objective nature and a buffered self, Miller presents Daoist alternatives such as the "subjectivity of nature" and the "porous self." I do not take a constructive approach as Miller does, but rather an anthropological approach in which I let my informants speak for themselves. In this chapter, I demonstrated how Chao-hwei also presented an alternative model to a buffered self and objective nature. She defended the subjectivity of nature and pushed not only beyond the buffered self but also beyond the porous self to the Buddhist concept of no self as a basis for the rights of the environment.

VI. CONCLUSION

John Locke's model of human equality carries with it a three-part creation-human-creator ontology, which transforms to a natural-human-supernatural ontology in secular form. While the distinction between animals and humans that the cosmology creates seems to be grounded more in imperial and scientific aspirations than in rigorous Biblical interpretations, the Christian model that Locke and his European contemporaries created is the inheritance of the modern liberal order. However, human-nature, human-supernatural, and nature-supernature divides did not exist in East Asia prior to contact with European colonial powers. Thus, the assignment of native Chinese teachings to the category of religion not only violated their own categories of knowledge but also isolated these teachings from the domain of power and restricted their power to make truth claims in the public sphere. These restrictions all occurred under a liberal order modeled on a European Christian cosmology. Chao-hwei presents an alternative to Locke's human equality that attempts to restructure these boundaries on Buddhist terms, reclaiming the equality of all sentient beings and the subjectivity of nature. However, in order to adapt these concepts to prevailing legal structures, she reformulates Buddhist teachings in the non-native liberal notions of "rights" and "environment." When her "rights of the environment" are understood to represent a Buddhist version of the "rights of nature," they reveal how she advocates employing the rights of nature to protect all sentient beings.

In 1996, the Taiwanese legislator Liu Ming-long (劉銘龍) campaigned on adding the rights of the environment to the Taiwanese constitution, but after his election, the idea gained little traction. Chao-hwei, LCA, and other environmental NGOs are aligned on implementing both animal rights and the rights of the environment. In her 2001 essay on the rights of the environment and animal rights, Chao-hwei argues that Taiwan must "solidify a larger social consensus" in order to succeed in putting the rights of the environment into law (2001: 316). For now, rather than constitutional amendments enshrining the rights of nature, LCA has employed fundamental laws to protect individual animals by way of Taiwan's 1998 Animal Protection Act and to protect species by way of the Taiwan's 1989 Wildlife Conservation Act. LCA and its allies also continue to amend these laws to expand protections for nonhuman animals, while at the same time supporting movements that protect both human and nonhuman animals. While these laws make inroads for nonhuman animals and the environment, they fall short of achieving parity with human rights, which are enshrined in Taiwan's constitution. Even if animal rights and rights of the environment receive constitutional protections, as long as human rights are enumerated separately, the constitution would need to harmonize human rights with animal rights and environmental rights as they stem from two incommensurable notions of equality. For example, the right of property granted to the people in Article 15 of the Constitution of the Republic of China would need to be reconceived to accommodate personhood or rights of nonhuman animals, species, and ecosystems.

The notion of earth jurisprudence, which includes rights of nature, is a legal philosophy that aims to take what Chao-hwei calls an eco-centric approach, one in which human laws are aligned with ecological principles.[43] Chao-hwei takes a pragmatic approach in which she accepts a legal notion of rights but rejects the cosmology that gave birth to the concept of rights. However, other platforms for earth jurisprudence exist. The Constitution of the Chinese Communist Party (CCP) in the People's Republic of China (PRC) enshrined the ecological civilization platform in 2012 (Wang et al. 2014: 37). This Marxist approach to earth jurisprudence aims to enter a third stage of civilization, one that supersedes industrial civilization, just as industrial civilization superseded agricultural civilization. The aim is a second Enlightenment that transcends the European Enlightenment. Part of the CCP's critique of the European Enlightenment is the anthropocentrism of a system of thinking that gave humans dominion over a nature from which they were separate. The CCP's antidote to this problem is a return to traditional Chinese teachings. Pan Yue, former vice minister of Environmental Protection in the PRC and a leading proponent of ecological civilization, frequently cites Buddhism's teaching "that all living things are equal" as a part of the ecological civilization platform.[44] The PRC's embrace of ecological civilization has earned it a global audience at the United Nations. While the PRC has virtually no laws that protect animals, Shih Chao-hwei, LCA, and Taiwan are forging a new path forward that may portend a new direction for a rising China that embraces ecological civilization and for those within China's increasingly global sphere of influence. Both rights of nature and ecological civilization offer new forms of earth jurisprudence that may accommodate and incorporate equality of life, so they invite legal innovation at scales ranging from Taiwan's island republic to the international forums of the United Nations.

NOTES

1. A shorter version of this chapter first appeared in the journal *ISLE: Interdisciplinary Studies in Literature and Environment* and reappears in the present volume with the original publisher's permission. The original citation is as follows: Jeffrey Nicolaisen, "Protecting Life in Taiwan: Can the Rights of Nature Protect All Sentient Beings?" *ISLE: Interdisciplinary Studies in Literature and Environment* 27, no. 3 (2020): 613–32.

2. I express my gratitude to the Charlotte W. Newcombe Foundation for the financial support of their Doctoral Dissertation Fellowship during the writing of this chapter, as well as to the Duke Global Asia Initiative for sponsoring the workshop from which this project grew, and to Prasenjit Duara, Hal Crimmel, Craig Kauffman, and Joni Adamson for generously providing feedback on the workshop draft of this article. I am grateful to Hwansoo Kim, Richard Jaffe, Ambika Aiyadurai, and Sean Riley for kindly providing helpful comments on later versions. I extend my gratitude to Mayfair Yang for her careful editing, patience, and editorial detail.

3. For Waldron's full argument about the basis of John Locke's equality, see Waldron chapter 3.

4. For Waldron's treatment of atheism, see Waldron chapter 8.

5. Chao-hwei consistently uses the term *zhongsheng pingdeng*, but as in the passage quoted below, she frequently uses *shengming* as the subject to which the concept of *zhongsheng pingdeng* applies. Other Buddhists in Taiwan use *shengming pingdeng* for the concept of *zhongsheng pingdeng*. "Equality of life" provides a somewhat more reader-friendly term in English, so I use that term.

6. The original Chinese was a slightly modified quote from Shih (2008: 60–61).

7. See Republic of China Ministry of Education, "Bentilun" [Ontology]. My translation.

8. The word *Chao-hwei* used for person here is *weige* (位格). This Chinese term is a technical term referring to person in the sense that the Christian Trinity is divided into three persons, different from the ordinary term for person *ren* (人).

9. My translation. A full English translation of *Fojiao Guifan Lunlixue* is also available: Shih (2014a).

10. Full quote: "A sage, having transcended self-views and self-love and attained the stage of no-self, of course can choose to abandon reasonable self-defense in order to protect others, but the ordinary person is limited by self-views and self-love . . . humans can expand and purify these strengths, perfect their enlightened nature, and be endowed with the Buddha's great wisdom, great compassion, and great heroism."

11. For an account of the transition from the concept of creation to that of nature in Europe, see Worster (1994: Chapter 1).

12. Article 1, UDHR, http://www.un.org/en/universal-declaration-human-rights/.

13. Shih 2001. An English translation is available in Shih (2014b). The article first appeared in *Huanzang Renwen Xuebao* in 2001.

14. Republic of China Ministry of Education 2015, "Tianfu." My translation.

15. My translation.

16. Chao-hwei clearly addresses the UDHR in this passage, and the term for "endowed" in the official Chinese version of the UDHR is *fuyou* (賦有). The *fu* (賦) means "to give," and is the same *fu* as in *tianfu quanli* (天賦權利) [God-given rights]. Thus, the connection between God-given, Creationism, and the language of the UDHR is much more explicit in Chinese than in English. See the Chinese version of the UDHR at http://www.un.org/zh/universal-declaration-human-rights/.

17. For Chao-hwei's arguments in this paragraph, see Shih (2003: 308).

18. For a review of environmental rights, see Boyd (2012).

19. This rights of nature concept is often credited to Berry. Berry uses the term *rights of natural modes of being*, rather than the term *rights of nature*. He also credits Christopher Stone and the Supreme Court Justice William Douglas as pioneering the concept of rights for nonhuman beings.

20. For a review of the rights of nature, see Boyd (2017). Its chapter 10 discusses Ecuador's constitution and chapter 11 discusses Bolivia's constitution.

21. See Hanyu dacidian bianji weiyuanhui (2001: 640). My translation.

22. Hanyu dacidian bianji weiyuanhui (2001: 640).

23. This paragraph draws from Chang (2019).

24. My translation.

25. The discussion of hun and po relies on Yu (1987). The quote from the Ji-yi is on page 393–394, including footnote 80. My translation.

26. The 3 hun and 7 po theory is well-known in Taiwan. Discussion of the hun and po theory in relation to nonhuman animals, however, is less common. Where explanations exist on popular websites in Taiwan, people tend to agree that nonhuman animals lack one hun.

27. For an account of the transformation of Wenchang, see Kleeman (1994).

28. *Mencius* (孟子), "Lilou" (離婁), Second part (下). My translation.

29. *Mencius*, "Jinxin" (盡心), First part (上). My translation.

30. *Zhuangzi*, "Mati" (馬蹄). Translation from Komjathy (2017: 31).

31. See Weller (2006: Chapter 2).

32. See Mackay (1895: 276–277).

33. For an article about the history of the Shen Pig Ritual, see Kuo (2018).

34. For a more detailed description of Qingshui Zushi, see Ministry of Interior National Religion Information Network (2019).

35. Chao-hwei's article imploring Chen not to participate in the ritual was originally published in the *Liberty Times*. See Shih (2013).

36. Harris (1997). The mention of the particular Indic candidate terms for nature is on pages 380–381.

37. For a more detailed discussion of the development of vegetarianism in China, see Grumbach (2005: 57–70).

38. Spiro (1966: 96 and 98) as quoted in Smith (1998: 281).

39. Interview of Kuo Cheng-tian, September 13, 2017.

40. See Jiaoyubu (2018: 36–38).

41. For a review, see Escobar (2010).

42. The highly regarded survey of Chinese religions by Goossaert and Palmer recognizes the problems with the category but opts to apply the term anyway. See Goossaert and Palmer (2011): introduction, especially 9–11. On the other hand, Lagerwey (2010) reasserted religion as a category for government Confucianism as a way to show that the Chinese government was not secular as Confucian scholars have tried to claim.

43. For a more detailed explanation, see Kauffman (2020).

44. See Wang, He, and Fan (2014: 54). See also Pan (2008: 30).

ENGLISH AND JAPANESE LANGUAGES

Aquinas, Thomas. 2004. "Summa Contra Gentiles." In *Animal Rights: A Historical Anthology*, translated by Anton C. Pegis, edited by Andrew Linzey and Paul Barry Clarke, pp. 7–12. New York: Columbia University Press.

Berry, Thomas. 1999. *The Great Work: Our Way into the Future*. New York: Bell Tower.

Boyd, David R. 2012. *The Environmental Rights Revolution: A Global Study of Constitutions, Human Rights, and the Environment.* Vancouver: University of British Columbia Press.

Boyd, David R. 2017. *The Rights of Nature: A Legal Revolution That Could Save the World.* Toronto: ECW Press.

Chang, Chia-ju. 2019. "Environing at the Margins: Huanjing as a Critical Practice." In *Chinese Environmental Humanities: Practices of Environing at the Margins,* edited by Chia-ju Chang, 1–32. Cham, Switzerland: Palgrave MacMillan.

Clough, David L. 2012. *On Animals: Systematic Theology,* vol. 1. London: Bloomsbury.

Coogan, Michael D., Marc Z. Brettler, Carol A. Newsom, and Pheme Perkins, eds. 2010. *The New Oxford Annotated Bible: New Revised Standard Version: With the Apocrypha: An Ecumenical Study Bible.* Fully Revised Fourth Edition. Oxford: Oxford University Press.

Descartes, René. 1990. "Animals as Automata." In *Animal Rights: A Historical Anthology,* edited by Paul A. B. Clarke and Andrew Linzey, translated by John Veitch, pp. 14–17. New York: Columbia University Press.

Escobar, Arturo. 2010. "Postconstructivist Political Ecologies." In *International Handbook of Environmental Sociology,* edited by Michael Redclift and Graham Woodgate, 91–105. Cheltenham, UK: Elgar.

Goossaert, Vincent, and David A. Palmer. 2011. *The Religious Question in Modern China.* Chicago: University of Chicago Press.

Gross, Aaron. 2014. *The Question of the Animal and Religion: Theoretical Stakes, Practical Implications.* New York: Columbia University Press.

Grumbach, Lisa. 2005. "Sacrifice and Salvation in Medieval Japan: Hunting and Meat in Religious Practice at Suwa Jinja." PhD Dissertation, Stanford University.

Harris, Ian. 1997. "Buddhism and the Discourse of Environmental Concern: Some Methodological Problems Considered." In *Buddhism and Ecology: The Interconnection of Dharma and Deeds,* edited by Mary Evelyn Tucker and Duncan Ryūken Williams, pp. 377–402. Cambridge: Harvard U Center for the Study of World Religions.

Holy Bible. King James Version, 2004. https://archive.org/details/KingJamesBib leKJVBiblePDF.

Kauffman, Craig M. 2020. "Managing People for the Benefit of the Land: Practicing Earth Jurisprudence in Te Urewera, New Zealand." *ISLE: Interdisciplinary Studies in Literature and Environment,* vol. 27, no. 3, pp. 578–595. doi: 10.1093/isle/isaa060.

Kleeman, Terry F. 1994. *A God's Own Tale: The Book of Transformations of Wenchang, the Divine Lord of Zitong.* Albany: State University of New York Press.

Komjathy, Louis. 2017. *Taming the Wild Horse: An Annotated Translation and Study of the Daoist Horse Taming Pictures.* New York: Columbia University Press.

Lagerwey, John. 2010. *China: A Religious State, Understanding China.* Hong Kong: Hong Kong University Press.

Locke, John. 1824. *An Essay Concerning Human Understanding.* Vol. II. New York: Valentine Seaman.

Mackay, George Leslie. 1895. *From Far Formosa: The Island, its People and Missions.* Fourth ed. New York: Fleming H. Revell Company.

Miller, James. 2017. *China's Green Religion: Daoism and the Quest for a Sustainable Future.* New York: Columbia University Press.

Orsi, Robert A. 2016. *History and Presence.* Cambridge, MA: The Belknap Press of Harvard University Press.

Pan, Yue. 2008. "Looking Forward to an Ecological Civilization." *China Today*, vol. 57, no. 11, pp. 29–30.

Rosenfield, Leonora Cohen. 1968. *From Beast-Machine to Man-Machine: Animal Soul in French Letters from Descartes to La Mettrie.* New York: Octagon Books.

Shih, Chao-hwei. 2014a. *Buddhist Normative Ethics.* Taoyuan: Dharma-Dhatu Publication.

Shih, Chao-hwei. 2014b. "Environmental Rights and Animal Rights: An Extension of the Concept of Human Rights and a Response to the Conviction of Protecting Life." In *Buddhist Normative Ethics*, edited by Chao-hwei Shih, translated by Chong Aik Lim, pp. 373–394. Taoyuan: Dharma-Dhatu Publication.

Shih, Chao-hwei. 2019. "An Exposition of the Buddhist Philosophy of Protecting Life and Animal Protection." In *Chinese Environmental Humanities: Practices of Environing at the Margins*, edited by Chia-ju Chang, translated by Jeffrey Nicolaisen, pp. 309–330. Cham: Palgrave Macmillan.

Shogakukan 小学館. "Shizen" 自然. Nihon kokugo daijiten 日本国語大辞典 [Japan National Language Dictionary]. Accessed March 30, 2014. https://japanknowledge.com/lib/display/?lid=200201e4120b4H4UG38C.

Singer, Peter. 1975. *Animal Liberation: A New Ethics for Our Treatment of Animals.* New York: New York Review.

Smith, Jonathan Z. 1998. "Religion, Religions, Religious." In *Critical Terms for Religious Studies*, edited by Mark C. Taylor, 269–284. Chicago: The University of Chicago Press.

Spiro, Melford E. 1966. "Religion: Problems of Definition and Explanation." In *Anthropological Approaches to the Study of Religion*, edited by Michael Banton. London: Tavistock.

Sterckx, Roel. 2002. *The Animal and the Daemon in Early China.* Albany: State University of New York Press.

Stone, Christopher D. 1972. "Should Trees Have Standing—Toward Legal Rights for Natural Objects." *Southern California Law Review*, no. 45, pp. 450–501.

Tse, Kuo 果澤. "Nothing Divine about Pig-fattening." *Taipei Times*, Feb 27, 2018. Accessed January 27, 2019. http://www.taipeitimes.com/News/editorials/archives/2018/02/27/2003688324/1.

Voas, David. 2013. "Religion, Religious Experience, and Education in Taiwan." In *Religious Experience in Contemporary Taiwan and China*, edited by Yen-zen Tsai, 187–212. Taipei: Chengchi University Press.

Waldron, Jeremy. 2002. *God, Locke, and Equality: Christian Foundations of John Locke's Political Thought.* Cambridge: Cambridge University Press.

Wang, Zhihe, Huili He, and Meijun Fan. 2014. "The Ecological Civilization Debate in China: The Role of Ecological Marxism and Constructive Postmodernism-Beyond the Predicament of Legislation." *Monthly Review*, vol. 66, no. 6, pp. 37–59.

Weller, Robert P. 2006. *Discovering Nature: Globalization and Environmental Culture in China and Taiwan.* Cambridge: Cambridge University Press.

Worster, Donald. 1994. *Nature's Economy: A History of Ecological Ideas.* Cambridge: Cambridge University Press.

Yu, Ying-Shih. 1987. "'O Soul, Come Back!' A Study in The Changing Conceptions of The Soul and Afterlife in Pre-Buddhist China." *Harvard Journal of Asiatic Studies*, vol. 47, no. 2, pp. 363–395. doi: 10.2307/2719187.

CHINESE LANGUAGE

Chao-hwei, Shih 釋昭慧. 2001. "Huanjingquan Yu Dongwuquan: 'Renquan' Guannian De Yanzhan Yu 'Husheng' Xinnian De Huiying 環境權與動物權——「人權」觀念的延展與「護生」信念的回應 [Environmental Rights and Animal Rights: An Extension of the Concept of 'Human Rights' and a Response from a Conviction in 'Protecting Life']." *Hsuanzang Renwen Xuebao* 玄奘人文學報 4: 17–34.

Chao-hwei, Shih 釋昭慧. 2003. *Fojiao Guifan Lunlixue* 佛教規範倫理學 [Buddhist Normative Ethics]. Taipei 臺北市: Dharma-Dhatu Publication 法界出版社.

Chao-hwei, Shih 釋昭慧. 2008. *Fojiao Houshe Lunlixue* 佛教後設倫理學 [Buddhist Meta-Ethics], *Yin Shun Daoshi Yuanji Sanzhounian Jinian* 印順導師圓寂三週年紀念. Taipei 臺北市: Dharma-Dhatu Publication 法界出版社.

Chao-hwei, Shih 釋昭慧. 2013. "Dao xia qingliu 'Zongtong zhu'! 刀下請留「總統豬」! [Put Down the Knife and Spare the President Pig!]." *Hongshi shuangyue kan* 弘誓雙月刊 (65): 4–6. doi: 10.29665/HS.200310.0002.

Chen, Chien-zong 陳建榮, and Huang Long-min 黃隆民. 2004. "Woguo guomin jiaoyu jieduan "Zongjiao yu xuexiao jiaoyu fenji" xiankuang de pingxi 我國國民教育階段「宗教與學校教育分際」現況的評析 [An Analysis on the Current Situation of "the Boundary between Religion and Schooling" in Elementary and Junior High Schooling]." *Taizhong shiyuan xuebao* 臺中師院學報 [Journal of Taichung Junior Teachers' College] 18 (1): 41–60.

Hanyu dacidian bianji weiyuanhui 漢語大辭典編輯委員會. 2001. "Huanjing" 環境. In *Hanyu dacidian* 漢語大辭典 4, 640. Shanghai: Hanyu dacidian chubanshe.

Jiaoyubu 教育部 [Ministry of Education]. 2018. *Shier-nian Guomin Jiben Jiaoyu Kecheng Wangyao: Guomin Zhongxiaoxue Ji Putongxing Goajidengxiao Yuwen Lingyu - Guoyuwen* 十二年國民基本教育課程綱要：國民中小學暨普通型高級等校語文領域－國語文 [Outline of the Twelve Year Basic Public Education Curriculum: Public Elementary and Junior High Schools and Ordinary Senior High Schools – Subject of Language and Literature— National Literature].

Ministry of Interior National Religion Information Network (*Neizhengbu quanguo zhongjiao zixuan wang* 內政部全國宗教資訊網). "清水祖師 (Qingshui Zushi)."

Accessed January 27, 2019. https://religion.moi.gov.tw/Knowledge/Content?ci=2&cid=285.

Pingguo ribao 蘋果日報. "Zongtong shenzhu mimi zai le 總統神豬秘密宰了 [The President Shen Pig Secretly Slaughtered]." August, 18, 2003, https://tw.appledaily.com/headline/daily/20030818/281164/.

Republic of China Ministry of Education 中華民國教育部. "Bentilun" 本體論 [Ontology]. *Jiaoyubu Chongbian Guoyu Cidian Xiudingben* 教育部重編國語辭典修訂本 [Ministry of Education Chinese Dictionary, Revised Edition]. 2015. http://dict.revised.moe.edu.tw/cgi-bin/cbdic/gsweb.cgi?o=dcbdic&searchid=Z00000017041.

Republic of China Ministry of Education 中華民國教育部. "Tianfu" 天賦 [God-given]. *Jiaoyubu Chongbian Guoyu Cidian Xiudingben* 教育部重編國語辭典修訂本 [Ministry of Education Chinese Dictionary, Revised Edition]. 2015. http://dict.revised.moe.edu.tw/cgi-bin/cbdic/gsweb.cgi?o=dcbdic&searchid=Z00000017041.

Singer, Peter 彼得·辛格. 1996. 動物解放 *Dong wu jie fang*. Translated by Meng Xiangsen 孟祥森, and Qian Yongxiang 錢永祥, Taipei 台北市: Guan huai sheng ming xie hui chu ban 關懷生命協會出版.

Chapter 2

Buddhist Environmentalism and Civic Engagement in Secular Shanghai

Mayfair Yang and Huang Weishan*

INTRODUCTION

This chapter examines the Buddhist Compassion Relief Tzu Chi Foundation[1] (慈濟基金會), a transnational Buddhist charity organization from Taiwan, and its environmental activities in the mega-city of Shanghai. Founded in Taiwan in 1966 by Dharma Master Cheng Yen (證嚴法師), Tzu Chi was a new Buddhist order of nuns, but its far-flung humanitarian activities are primarily managed by lay Buddhist members.[2] Tzu Chi offers a new kind of Buddhist practice that is skilfully adapted to modern secular society. It emphasizes social activism to provide disaster, poverty, and medical relief to human suffering, rather than seeking personal religious salvation through scriptural study, meditation, or ritual performance. We show how this Buddhist organization conducts its religio-environmentalist activities within a largely secular society by creating social niches focused on community voluntarism and grassroots education for spiritual and ethical reform. The fieldwork and interviews with Tzu Chi members were carried out by Huang Weishan between 2010 and 2015 in Shanghai (in Xuhui, Changning, and Pudong Districts) and in Beijing, in trips lasting one to three months each. We begin with an ethnographic vignette from Huang's fieldnotes on a weekday in 2011:

I attended a Tzu Chi Foundation public event in the center of Shanghai. I was welcomed by Tzu Chi sisters in the lobby of their business building and promptly instructed to use the stairs instead of the lift. Once I reached the door of the fifth floor, I had to remove my shoes, place them into a reusable shoe bag, and put on ecofriendly indoor socks. Like many others, I was enjoined to avoid using plastic bags or disposable bottles. We quickly washed our hands to save water before we entered the main hall. I sat on a low cushioned taboret with more than 200 practitioners, waiting silently. Vegetarian dinner boxes were served with ecofriendly tableware. After a simple meal, we watched an educational video about climate change and how humans could alter the path of global warming. After the video presentation, several commissioners dressed in blue uniforms took turns giving PowerPoint presentations on teachings from one of the main Tzu Chi environmental protection missions. Most of the participants were new to the group and were accompanied by senior Tzu Chi commissioners. Some were familiar with the Tzu Chi mission and would echo a speaker's talk with shouts of Tzu Chi slogans. The program ended with sign language singing and rites of repentance for the sin of being materialistic inhabitants of the earth.

Similar scenarios transpired at other Tzu Chi District Centers in Shanghai's Pudong, Putuo, Songjiang, and Beijing. Through the organization's educational videos, participants learn of actions taken against climate change on the personal, household, city, and global levels. Converted members thus become involved in the global project of "saving the Earth," while also advancing their own spiritual growth.

This chapter focuses attention on four questions regarding how Tzu Chi, a major socially engaged Buddhist organization, manages its environmental activities in China's largest city. First, how does Tzu Chi break out from its parochial Taiwan origins and develop a global perspective? Second, how does Tzu Chi balance the cultivation of the self with the religious transformation of the world, and integrate the private and public spaces of its religio-environmentalist activities? Third, how does Tzu Chi introduce religio-environmental innovations that unsettle hegemonic ontologies of modern secular urban society? Fourth, how does Tzu Chi negotiate the shifting boundaries between religious and secular life, and manage to gain a footing and survive in the highly secular society of Shanghai? These four dimensions are not separate, but interconnected in Tzu Chi's religious environmentalist work, whose success is predicated on the skillful negotiation of all four aspects of its socially engaged Buddhism.

Shanghai has a large population of Taiwan businessmen, entrepreneurs, professionals, and students working and residing long-term, often with their families. The Tzu Chi Foundation was brought to Shanghai by transnational Taiwanese entrepreneurs in the early 1990s. As in Taiwan and elsewhere, middle-class and middle-aged women figure prominently in Tzu Chi membership and activism, so the wives of Taiwanese businessmen were active proselytizers. The Tzu Chi organization engages in eight mission activities (八大志業 or 八大法印): (1) charity and poverty relief; (2) medical care; (3) education; (4) culture; (5) international disaster relief; (6) bone marrow donation; (7) environmental protection; and (8) community volunteer. At the time of writing, the Tzu Chi Foundation is the only Chinese Buddhist organization that has made environmentalism one of its official organizational missions. An official mission status guarantees the allocation of resources, annual goals for implementation, as well as internal project reviews.

I. FROM HUMBLE BEGINNINGS TO TRANSNATIONAL BUDDHIST OUTREACH

Buddhism is one of the world's foremost global cosmopolitan religions, having crossed countless ethnic, linguistic, political, religious, and national boundaries ever since it started leaving its cradle in northern India over two millennia ago. Whether transmitting Buddhist teachings by land or by sea, and in modern times by air, Buddhism had to balance its universal teachings while also embedding and adapting itself into the local religious and political landscape. After Mahayana Buddhism moved into China from India in the second century CE, devout Chinese converts traveled between the two lands to collect scriptures and study the teachings, while systematically translating Buddhist scriptures into Chinese. From China, Mahayana Buddhism was transmitted to Japan and Korea, where new translation projects took place. Theravada Buddhism spread to Sri Lanka and Southeast Asia, except Vietnam. In China, it learned to subsume itself to the temporal power of political authority, which was stronger than in India, and incorporated Confucian filial piety and Daoist thought into its teachings. Over the centuries, Buddhism's important teachings of *karma* (*ye* 業 or *yuan* 緣) or the law of cause and effect, merit accumulation, and compassion became indistinguishable from basic Chinese cultural values.

In his work, "How Climate Change Might Save the World," Ulrich Beck reiterates his critique of "methodological nationalism" (2016), inherited from the nineteenth century. For Beck, in the age of the Anthropocene, this nineteenth-century knowledge system of basing the world around the concerns of the nation must be substituted by a new "methodological cosmopolitanism," which would base the nation around the "world at risk." While a majority of theorists assume that the space of transnational society convincingly develops only as a consequence of deliberate action, Beck's theory of the world risk society argues that global problems are forcing us together involuntarily and out of dire necessity. A global problem cannot be solved by individual nations. Modern nation-states jealously guard their sovereign borders and access to natural resources for their own citizens. Nation-states also engage in rivalry over GDP, commodity production and trade, and global cultural prestige, so their efforts are usually restricted to benefiting their own citizens. Nation-state rivalry and lack of cooperation were demonstrated many times at international meetings to tackle global climate change. Third World countries thought advanced industrial nations should reduce carbon footprints more radically, since they were the ones who caused most of the problems. Wealthy nations thought it unfair to allow Third World countries to pollute the world in their late industrialization drives. However, global problems such as climate change and COVID-19 require a global outlook, a form of cosmopolitanism to overcome such global risks. Thus, global climate change and other environmental degradations that threaten life on this planet have forced us humans into thinking politically in order to change our destructive modern culture and social order.

Given the self-interests of nation-states and their dragging of their feet to address climate change, the world must increasingly also rely on global and cosmopolitan local civil societies that transect nation-states, to carry out efforts to help our planet. Our consciousness of a risk society of global climate change and the COVID-19 pandemic push our scholarship to depart from the old "methodological nationalism". With its heritage of cosmopolitan outlooks developed in ancient missionizing efforts, Buddhism has many strengths to tackle global environmental problems to relieve human suffering and administer spiritual salvation. When an ideal Buddhist looks at another person, he or she should not discriminate based on nationality or race, class or status position, gender, or age, for all categories of persons experience suffering and need help in relief from suffering. As Wei Dedong in chapter 6 also shows, Buddhists are supposed to feel compassion for the suffering of all life forms, such as fish and turtles, and care not just for human well-being.

More than other Taiwanese Buddhist organizations, Tzu Chi has built up an impressive network of transnational organizations across all continents except Antarctica (Huang, J 2013). It started out as the brave humanitarian initiative of a humble Taiwanese nun with the help of thirty housewives in Hualien, a backwater of Taiwan. Today, it is one of the largest organized Buddhist transnational organizations in the world today, with over ten million members around the globe. Tzu Chi Foundation boasts more than 445 Tzu Chi offices in forty-eight countries, including five countries in Africa, thirteen in Latin America, nine in Europe, fifteen in Asia, and Australia and New Zealand (Huang, J 2013; Huang, W 2011).[3] There are a total of 77,621 Tzu Chi certified "Lay Commissioners" and "Faith Corps"[4] members globally in thirty-four regions of the world (Tzu Chi Fact Sheet 2012). Moreover, there are a further 76,219 registered Tzu Chi volunteers in over seventeen countries. Just as early Buddhism adapted itself to local cultures and political authorities, today, Tzu Chi has also learned to carefully traverse the divide and tensions between religious and secular life in the countries it enters. It studiously avoids entanglement in politics and does not make political statements. It also refrains from invoking any nativist Taiwan political identity, even while so many of its Commissioners, Faith Corps, and volunteers around the globe come from Taiwan.

In addition to its cosmopolitan tendencies, Buddhist teachings provide an extremely fertile field for modern environmentalist teachings and ethics. Within its vast scriptural corpus, there is very little hard division between nature and culture, as Buddhist teachings emphasize that all beings, whether human or nonhuman, have basic Buddha nature. While in its long history, interpretations of Buddhist writings and actual Buddhist practices could be said to have an anthropocentric tendency, nevertheless, there is vast potential in this written corpus for modern environmental reinterpretations, as Jeffrey Nicolaisen in chapter 1 has already demonstrated with the work of Taiwan Buddhist philosopher Dharma Master Chao-hwei. In this chapter, we will examine how a living tradition of Buddhism, in the form of Tzu Chi Foundation, continues to adapt Buddhist teachings and practice to new socio-historical contexts, in this case, the highly secular society of contemporary Shanghai, and introduces philosophical and ontological innovations to assist in adapting Buddhist teachings to address the current environmental crisis.

II. PRIVATE AND PUBLIC SPACES: SELF AND WORLD

Like most religious cultures, Tzu Chi prescribes the ideal religious comportment for its members both inside the home and outside in public spaces of society. Tzu Chi's green missions or environmental actions lie at the heart of its interactions with secular society.

The Private Domestic Sphere

In the private sphere of the home, these missions adhere to a series of four disciplines supporting the Buddhist task of spiritual self-cultivation (修行 *xiuxing*). These disciplines ingeniously embed the self as an object of religio-environmental self-discipline within both the modern family and urban technological context and secular framework for positive social actions.

Home Temperatures

Huang Weishan's fieldwork in Shanghai during the cold winter of 2010 still brings back memories of attending spiritual teaching sessions in a Tzu Chi sister's frigid living room. When she visited Tzu Chi commissioners at their homes in Shanghai, none of the hostesses would turn their heaters on for reasons of resource conservation. They cautiously followed the instructions of the Tzu Chi Foundation on how to live a green and prudent lifestyle. Water conservation was also considered to be cultivating an attitude of respect for a water source and extending the life of water. Tzu Chi practitioners carefully collected and reused their water for rinsing vegetables, dishes, and even laundry. Water that had been used for bathing was stored and used later for flushing the toilet.

Women and Vegetarianism

Housewives have traditionally been the backbone of Tzu Chi manpower in many countries. Tzu Chi has a rather conservative attitude toward women's roles in marriage and the family. Being wise but submissive wives is often more important than maintaining egalitarian partnerships in marriage. Nevertheless, housewives have an important role to play both inside and outside the home. Many female Commissioners balance their lives by serving as community volunteers outside the home during the day, and gradually convincing their family members to join Tzu Chi missions while at home. The vast majority of Tzu Chi housewives are practicing vegetarians themselves, although their family members may not be. They use their traditional gender roles to exert influence on their husbands, children, and elderly parents or in-laws. The housewives are taught to ingeniously invite their husbands and children to enjoy vegetarian dishes and become habituated to them. In Shanghai, with the encouragement of their wives, male entrepreneurs have increasingly joined the community services and disaster relief efforts, eventually taking up local leadership in some cases.

Ethical Eating

Ethical eating is a key practice that Commissioners leverage in their private spheres as household managers. Today, ethical eating is no longer confined to a vegetarian diet as in traditional Buddhism. Ethical eating now means consuming locally produced food and avoiding gourmet food transported over great distances. In Shanghai, urban residents, sometimes spoke of the challenges of this discipline, since most of the city's available food products are shipped from outside the urban area. Finally, ethical eating is a symbolic act of repentance during an era marked by natural disasters such as floods, earthquakes, and the COVID pandemic crisis. Members are instructed to repent and pray for a disaster-free world by taking up vegetarian diets. In February 2020, Tzu Chi headquarters kicked off a campaign to repent for the crisis of the spread of COVID-19. The "Global Prayer and Vegetarian Fast," aimed to eliminate the epidemic through daily global collective prayer at 1:30 pm, accompanied by ethical eating.

Recycling

Reusing or recycling waste is another important household task for practitioners, with a thirty-year history among the many activities of Tzu Chi

Foundation. On August 23, 1990, Master Cheng Yen made a speech at a school in Taichung City, Taiwan, in which she called upon everyone to make Taiwan a beautiful island by working with the government to sort out garbage and recycle them to new uses. When the audience of one thousand clapped vigorously after her speech, she responded, "Use your applauding hands to sort out garbage!" Thus, began various island-wide campaigns called "Preserve a Pure Land on Earth" to plant trees, recycle, clean up beaches, streets, and river banks, and to organize lectures and concerts to raise people's consciousness about the environment. In 1998, the total amount of waste disposal in Taiwan was 8.8 million metric tons. That number decreased to 7.8 million in 2000, and then to only 5 million metric tons in 2006. Meanwhile, the amount of recycled garbage in Taiwan increased by nearly seventeen times, from 129,000 metric tons in 1998, to 2,188,000 metric tons in 2006.[5] All these recycling efforts meant that the Taiwan government was able to cut down its planned construction of garbage incinerators from thirty-six to only twenty-four, greatly saving on government funds and energy used in burning. Outside Taiwan, Tzu Chi now operates 5,462 recycling centers around the world (figure 2.1).

In Shanghai, recycling was first mandated by Chinese law on January 31, 2019.[6] When Mayfair Yang visited Shanghai in December of that year, she

Figure 2.1 Sister Mei demonstrates her banner for the recycling campaign in Baoshan District, Shanghai, 2011. Photo by Huang Weishan.

noticed brand new recycling bins installed everywhere in neighborhoods, with government wall posters and public messages on television instructing people how to sort their garbage. However, the Tzu Chi Foundation had already established and operated recycling centers in Shanghai since 2010. During interviews in Shanghai, Huang Weishan found that many Tzu Chi members quoted the maxim of "turning waste into gold" to describe their efforts of giving a "second life" to unwanted items at home. Tzu Chi devotees carefully collected, washed, and sorted through different kinds of used cans, bottles, and paper products in their homes and took them to neighboring recycling centers. They even saved organic wastes to feed their plants. These were time-consuming processes, and Tzu Chi members seemed to be especially vigilant, disciplined, and enthusiastic about it, given extra motivation, no doubt, by their Buddhist faith. In contrast, a cursory look at recycling bins in public spaces showed that non-Tzu Chi people tended to carelessly throw in garbage or did not bother to sort the different kinds of recyclables.

The mainstreaming of the green mission among Buddhist communities in China was not successful until the implementation of the government's top-down policies in 2019. On January 31, 2019, the Shanghai Municipal People's Assembly passed "The Regulations of Shanghai Municipality on Municipal Solid Waste Management," dealing with domestic household wastes, and not industrial wastes.[7] All legal religious, educational, and media organizations, including Buddhist temples and monastic spaces, were mobilized to serve as social units for educating the public about the implementation of new regulations on waste management, especially recycling. In a matter of a few weeks, environmental protection and recycling education became key concerns for all of Shanghai religious communities. Monastic clergies and temple volunteers were mobilized to teach lay believers on topics such as recycling and other forms of waste management, and prepare people for implementing the policy on July 1, 2019. This regulation gave a boost and lended further legitimacy to the ongoing recycling efforts by Tzu Chi in Shanghai.

If the self is a product of *habitus* or repeated customary acts that reproduce and reinforce one's sense of self and self-identity, and prevent it from the degeneration of entropy, then Tzu Chi members follow mundane daily routines that both remind them of their special identities whose activities align with the protection of the earth, as well as provide a guide or means for living a Buddhist life of self-cultivation. The task of constructing a religious self-identity begins with considering the kind of food that one puts inside one's body: vegetarian and locally and sustainably produced simple non-luxury foods, that are not shipped over great distances. Tzu Chi devotees are encouraged to extend this daily regimen to their inner family circle, so as to persuade and realign their selves and family to vegetarianism and ethical eating. Next, the responsibility of recycling brings family members and neighbors together in a common task of sorting and transporting their family consumption items to local recycling centers, and repurposing those items.

Volunteer mobilization in public space

Tzu Chi volunteers mobilize collectively in public places for missions in poverty aid, disaster relief, environmental protection, cultural education, and medical fields. In contrast with the activities of public monasteries, Tzu Chi's volunteer mobilization efforts can be understood as expressions of moral reformation in secular settings. This understanding is especially helpful for analyzing the very secular context of religious revival in contemporary China. In the following section, we will describe mobilization efforts for environmental protection taking place in Shanghai, Beijing, and Taipei, such as street cleaning, recycling, and the educational efforts in religious ethics at the local district centers and companies. A central belief for Tzu Chi practitioners is that "practice" (*xing* 行) is the outward manifestation of "belief" or "faith" (*xin* 信). For Tzu Chi members, belief is very important to them, as it is their spiritual commitment to the religious community, to Master Cheng Yen, and to their whole karmic cycle of existence and rebirth. However, belief must be expressed or proven in outward conduct, by their voluntary contributions to causes deemed important in Tzu Chi culture.

Both in Shanghai and Beijing, Huang Weishan took part in green missions at various recycling booths and stops, and interviewed the local organizers. Both Taiwanese and Chinese Tzu Chi Commissioners distributed their volunteer manpower to target local "urban and suburban districts" (市區), an urban administrative unit in Chinese cities that oversees urban functions such parking and traffic, business permits, policing and courts, and so on. For example, in Xuhui District in Shanghai, with a total population of 1.08 million people, Tzu Chi organizers held recycling events on Saturday mornings and partnered up with several neighborhood residential committees (居民委員會), which are sub-administrative units under the districts. Most of the practitioners were young Chinese trainees who only had every other Saturday off from work. The partnership with the neighborhood committee provided a stable weekly schedule and audience for both Tzu Chi's effort to recruit new members as well as its waste collection activities, where it could show people the value of group efforts in environmental protection. With rising awareness of pollution and climate change, the Tzu Chi Foundation of China has transformed the Xuhui District recycling stations into an educational center promoting sustainable development and an environmentalist urban consciousness. It also actively promoted sustainable lifestyles and the zero-waste movement. Through Tzu Chi's public educational efforts, recycling, which used to be considered purely labor, has gradually started to be considered an admirable "lifestyle," packaged to fit young urban residents' tastes.

In Huangpu District, Shanghai, volunteers held their recycling events in the evenings on Xizhang South Road, the downtown area of the district, because most of the volunteers were young and had day jobs. While passersby were rushing home around 7:00 pm, Tzu Chi volunteers were busy recycling on the side of the road. This "evening environmental protection" (夜間環保) has gradually become a routine adapted to young Tzu Chi members' schedules. In Baoshan District, Shanghai, the company site of a Taiwanese entrepreneur was turned into a local center for recycling and education in 2010. Local community residents easily spotted and joined Tzu Chi gatherings on week nights in this office-based center because of the music and crowds. In 2012, Huang Weishan found about sixty people of all ages singing and dancing at this private office turned into a public recycling center. In interviews between 2010 to 2014, she found that most of the local residents had positive reactions and were attracted to the Tzu Chi group because of its green mission.

In 2014, a Beijing training camp was established to promote the use of self-made fruit waste enzymes as an organic replacement for harsh chemical detergents and household cleaners. The camp was hugely successful and the Tzu Chi Center was later established to assume the mantle of carrying on green education programs on fruit waste enzymes. Located in a suburban area a one-hour subway ride away from Beijing city, this center is one of the largest Tzu Chi district hubs. The center also serves as a regional base for retreat camps, at one time hosting 500 trainees who traveled from northeastern China, with the permission of local officials, to work closely with the nearby Neighborhood Residential Committees. Training programs take place once or twice per year, depending on annual numbers of newly converted members. Training programs cover Tzu Chi's eight missions but also highlight the green lifestyle.

Two Tzu Chi *Still Thought Bookstores* in Beijing, named after the main ritual hall of Tzu Chi's sacred headquarters in Hualien, Taiwan, serve as Tzu Chi centers for small-scale cultural events and green missions. In one public street cleaning campaign Huang attended in Qianmen, West Street in 2014, more than thirty trainees joined the event, attracting the attention of commuters and tourists alike. The cleaning was followed with video courses that were open to the public, on Tzu Chi's green mission inside the bookstores. One of the *Still Thought Bookstores*, hosted weekly study groups organized by an established Chinese entrepreneur whose enthusiasm for environmental protection led to his becoming ordained as a Buddhist monk by Master Cheng Yen in 2014. Green missions have continued to be the most successful of the organization's efforts to encourage volunteerism in Shanghai and Beijing, as they appeal to both the urban newly rich middle class, to Chinese entrepreneurs, and to young people in their teens through thirties.

Finally, the Neihu Environmental Protection Station in Taipei has developed its own unique cultural style that one does not find in Mainland China, where environmental enthusiasts tend to be young. The Neihu Recycling Station, like many other recycling stations in Taiwan, is a place for elderly, retired Tzu Chi volunteers to spend their days as full-time volunteers working on the detail-oriented task of classifying all the recycling materials. The station functions not only as a secular recycling station for volunteers but also as a religious site for worshipping. During Huang's visit in 2016, volunteers conducted hourly prayers while prayer music was broadcasted from loud speakers. In the recycling process, the Tzu Chi self is proudly displayed to the public, while at the same time, contributing to building up group religious identity and religious recruitment efforts.

Building a Neighborhood-Based Urban Life

Tzuchians once focused on social engagement in rural areas. More recently since 2014, Tzu Chi has started recentering its green missions in urban residential communities in both Shanghai and Beijing. This has created new urban social spaces to recruit urban Tzu Chi members. By using secular language and engaging in environmental protection, Tzu Chi members are able to act on the ethic of reformed Buddhism in highly regularized and controlled urban spaces. This structural adaptation corresponded with a broader shift in social identity taking place in recent years in urban Shanghai, resulting from governmental policies shifting urban governance and administration from the old "work units" (*danwei* 單位) to accommodating local territorial and "residential communities" (社區). With the permission of local authorities, Tzu Chi practices are implemented in residential communities and senior centers. Members now more often identify themselves by territorial and community-focused district affiliations, rather than their place of work (figure 2.2).

By responding to the institutional project of integrating service with local urban communities, the Tzu Chi movement has become highly localized via its increasing local membership and grassroots spaces that their residential communities provide for the realization of socially engaged Buddhism. Although Tzu Chi is a religious organization, cooperation between Tzu Chi members and the local government, the Residential Committee, is possible, in part because Tzu Chi challenges the social norms of ethical values, but not Chinese government or Party policies, nor the Chinese political system. Nevertheless, while observing successful projects in residential communities in Shanghai and Beijing, Commissioners did express some concerns. They worried that one day, a change of policy at the national and local levels of the

Figure 2.2 A recycling activity in Xuhui District, Shanghai, 2016. Photo by Huang Weishan.

government might suddenly lay waste their hard-won environmental efforts. For the moment at least, their cultivation of "the community Dharma field" (*shequ futian* 社區福田) continues (Huang 2018).

III. INTRODUCING A NEW RELIGIO-ENVIRONMENTAL ONTOLOGY

This section will address how Tzu Chi expounds upon ancient Buddhist ontology and its notion of human existence as part of the ever-changing cycles of the birth, life, death, and rebirth of "sentient beings" (*zhongsheng* 衆生). Sentient beings are those life forms that have consciousness and feelings, and experience suffering, including insects, fish, animals, and humans, but usually not plant life. Although plants are not regarded as sentient, and are eaten by Buddhist vegetarians, they must also be cared for because they are part of the earth. We examine how Tzu Chi adapts ancient Buddhist doctrines to the task of changing our modern anthropocentric ontology that promotes rampant industrial production and consumerism.

The Impermanence and Surface Appearance of Things in the Cosmos

Tzu Chi teachings follow traditional Buddhism's understanding that all phenomena in the cosmos are impermanent, including both physical matter and the lives of all sentient beings. Master Cheng Yen once said:

我們身處在這個地球上，所有的一切都離不開「三理四相」，物質有成、住、壞、空；身體有生、老、病、死；心理有生、住、異、滅，然而這一切，最終還是回歸於「眾生共業」．

We are positioned to live our lives on this planet, and [must remember that] all things never diverge from the principle of "the three rationalities and four appearances." The rationality of all physical matter goes through the epochs (*kalpas*) of formation, existence, destruction, and Void. The rationality of all living bodies goes through the epochs of birth, aging, sickness, and death. The rationality of all consciousness goes through the epochs of genesis, existence, change, and extinction. Thus, all phenomena will in the end return to the "collective *karma* of sentient beings."[8]

As one acolyte, Hsiao Chiu-ling, explained on a Tzu Chi webpage, the point that Master Cheng was making was that we should not be misled by the outward appearances (相) of phenomena, nor get attached to them, for all things follow the essential principle that they will eventually decay into nothingness or enter the void (空). However, human actions create imbalances in the cosmic timetable, and will hasten the movement toward destruction of the cosmos.

大地生（成），生態具足，四季順暢運轉，這是（住）時。但是當人心貪欲產生濁氣，不斷向大地攫取，大地就受災，被破壞了；破壞（住）劫，進入到（壞）劫。長時間破壞，到了最嚴重的時候，完全毀滅，那就是（空）。

The Great Earth is "formed," with an abundant and balanced ecology and the four seasons smoothly following its proper rotations. This is the "existence" epoch. However, when humanity's greed and desire give rise to a foul odor, and humanity ceaselessly seizes and robs from the Great Earth, the Earth will experience calamities, its "existence" will decay, and it will enter into the epoch of "destruction." As the Earth continues to be ravaged for a long time, when this process reaches a most serious phase, it will be completely destroyed, and it will enter into the state of the Void.

—Hsiao Chiu-ling[9]

Here, we can see that, although Tzu Chi doctrine teaches that all physical and living phenomena on Earth are subject to ceaseless change, decline, and death, human actions may speed up the path toward destruction and extinction. The idea here is that in human activities, there should be balance and moderation, so that cosmic forces may go about their natural cycles and timely transformations, just as the transitional flow of the seasons.

The Tzu Chi author Hsiao then provides a historical example of the consequences of human greed and the plunder of the earth. She recounts the story of a king of the Pagan Kingdom (849–1297) in Burma who was a devout Buddhist. He was greedy in accumulating merit by building vast numbers of stupas throughout the kingdom. Today, Burma still has 10,000 ancient Buddhist stupas in various stages of repair. The king not only depleted the state treasury but also cut down vast swaths of forests for wood to burn in making the bricks for his construction projects. The depletion of trees caused the abundant ground water to gradually dry up. Thus, wrote Hsiao, "When humanity disrupts the natural cycles of the environment, it will propel us from the epoch of 'existence' to that of 'destruction,' and in the end move us rapidly into the state of the 'Void.' "[10] Here, we can see that contemporary Tzu Chi teachings diverge from traditional Buddhism, which tended to urge transcendence and withdrawal from worldly life, and encourage the diverting of material wealth into religious expenditures to glorify Buddhist symbols and structures. Instead, Tzu Chi enjoins its members to focus on this worldly concern of living modestly and frugally in harmony with the natural environment. For Tzu Chi teachings in the urban risk society of today, even ancient Pagan's building of Buddhist stupas and temples that promoted the faith was wrong, for it depleted the precious forests.

Karmic Relationships between Humans, Nonhuman Life, and the Planet

Tzu Chi environmental ethics places great importance on the human-animal and the human-ecosystem relationship. Its discourse of vegetarian dietary practices is directed by Buddhist ethics governing human interactions with the natural world and its sentient beings and organic systems. Public testimonies given by Tzu Chi members converting to a vegetarian diet typically cite not only the Buddhist commandment not to kill but also reveal an overarching focus on practicing compassion for the earth and its living beings. This sense of compassion is also visible in Buddhist interpretations of natural disasters, which are often empathetically understood as nature fighting back against the bad deeds of human beings. Thus, Tzu Chi also expounds on the human-*karmic* relationship with the natural world. In keeping with its this worldly focus, Tzu Chi discourse emphasizes that human destructiveness and hatred toward other sentient beings will accumulate negative "collective *karma*" (*gongye* 共業) for humanity as a whole, and be repaid in this life, in the form of natural disasters and human suffering from hunger and war. In order to take control of their "impermanent" lives and the health and survival of planet earth, Tzu Chi members need to take actions and repair our collective human *karma*.

In contrast to traditional Buddhism, Tzu Chi members talk more often about "collective *karma*" than about "individual *karma*" (獨業). In telling the aforementioned story about the king of Pagan, the Tzu Chi writer Hsiao

deplores the king's indulgent wasting of wealth and natural resources, saying that, centuries later, a historical earthquake toppled almost 500 stupas,
destroying so many in one fell swoop, rendering all the human efforts of
building them meaningless and worthless. For Hsiao, the earthquake was a
result of the "imbalance" (偏差) in the cosmos produced by the king's desires
and his greed in merit accumulation, which produced negative collective
karma for the people several generations later.[11] Similarly, at a meeting held
at Tzu Chi Headquarters in Hualien, Taiwan in 2009, Huang Weishan heard
Master Cheng Yen specifically address the U.S.-based Tzu Chi Commissioners gathered there, amidst Commissioners from many other countries, "You
American volunteers should put out more effort [in Tzu Chi charity work].
Your government sent troops to fight in Iraq, that's why you got several tor-
nados and flooding events." Thus, whether human beings can live together
peacefully and harmoniously in contentment, or must suffer from depriva-
tion and conflict, or from natural disasters and environmental degradation,
depends on the past collective actions of a group, a national population, or
humanity as a whole.

Self-Cultivation as Environmentalist Action

Tzu Chi's basic ontological claim about human nature follows traditional
Buddhism in observing that everyone is born with a "Buddha nature" (佛性)
and has the potential to achieve certain levels of Enlightenment and even Buddhahood through "self-cultivation" (修行 *xiuxing*). Although human beings
are born with a Buddha nature, this reality may be hidden. Self-cultivation
is a method of making oneself aware of one's true Buddha nature. Tzu Chi
counsels its followers to seek transcendence from the cycle of endless rebirths
and lives of suffering, but in keeping with its secular proclivities, generally
refrains from going into details about past and future lives. For Tzu Chi, this
self-cultivation differs from the history of Buddhism's emphasis on scripture
chanting and meditation, and takes the form of secular action in this life, in
and for the sentient beings of this world. "Learning and following Buddha's
path" (學佛 *xuefo*) is a way to unveil the true Buddha nature, which lies in
showing "compassion" (慈悲) for others. The Buddhist value of a life of
"wisdom" (慧命 *huiming*) in Tzu Chi discourse becomes the wisdom to see
and empathize with the suffering of all sentient beings in their deteriorating
environment. Thus, environmentalist action is linked to the most important
Buddhist field of action, self-cultivation, which, in turn, becomes a religious
rationale for environmental ethics.

Master Cheng Yen directly makes this connection between traditional Buddhist self-cultivation to attain Enlightenment with the health and welfare of
planet earth. In the world of the "ten directions" (十方 or the whole universe),

earth is the only place where one can cultivate oneself toward Buddhahood. Master Cheng Yen said, "Buddha has taught people that human beings cannot become Buddha in heaven nor in hell. The only place for cultivation is the earth. Therefore, if human beings destroy the Earth, we destroy our lives of wisdom." Thus, if earth is destroyed, then the only place in the cosmos where sentient beings can cultivate themselves and attain Enlightenment will disappear. This rationale for saving earth is parallel to traditional Buddhist teachings that, although all sentient beings have the potential to become Buddhas, they need to wait until they are reborn into a human being, in order to cultivate and attain Enlightenment. However, Tzu Chi adds an extra new modern environmental layer to ancient Buddhist doctrines, while at the same time infusing religious and cosmic insights into secular environmentalism. Tzu Chi is warning that, unless we save the earth, all sentient beings will forever remain trapped in the darkness of endless and pointless cycles of birth, suffering, and rebirth.

The Tzu Chi concern for protecting planet earth, the only physical place where human beings and other sentient beings can survive in this life, is not purely about biological survival. There is also an altruistic tinge, in that compassion and wisdom connect to a globally interconnective consciousness that could eventually enable humans to save the planet together. By saving the earth, one is essentially also saving others, and the conditions for everyone to attain Enlightenment. In this way, environmental actions are part of a broader mission of compassion for both the global human community and the earth's living beings. In public speeches, Master Cheng Yen calls Tzu Chi environmentalist volunteers in any country, "spiritual farmers nurturing the Earth," alternately referring to them as "Environmental Bodhisattvas" (環保菩薩), "Purified Bodhisattvas" (清淨菩薩), or "Bodhisattvas who Embrace the Earth" (擁抱地球的菩薩).

The Human Relationship with Inanimate Synthetic Objects

While many Asian religions provide explanations about humanity's relationship with the natural world, Tzu Chi extends these teachings to cover the built environment or synthetic objects in the secular world. Master Cheng Yen has pioneered a new mode of ontological thinking about the inorganic, inanimate world of synthetic human-made objects. Since the Industrial Revolution and the post–World War II global consumer culture, our planet has become awash with synthetic human-made materials and objects. Mountains of garbage dot the landscapes of every country and are buried under the surface of the earth. Huge floating islands of plastic and other garbage can be found on the surface as well as deep within the Pacific Ocean and remote seas, killing the marine wildlife or seabirds who mistake them for food. Cheng Yen makes an ontological breakthrough in getting her followers to talk about and treat these inanimate synthetic objects produced by human beings as living objects, or what she calls "life of material objects" (物命 *wuming*). In investing them with life force, Cheng Yen is not only extending the usefulness and socio-economic value of these material objects, but also reanimating their lifeless forms so that humans may respect and even feel compassion for objects and things and extend their "life."

The ancient Chinese notion of *wuming* in the Chinese classics was on non-human life, such as in Wang Chong's Han dynasty composition, "On Measuring: The Longevity of Primary Breath." A passage in this text compares human lifespans with the nonhuman lives of plants and animals, which have their different natural allotted timespans.

百歲之壽，蓋人年之正數也，猶物至秋而死，物命之正期也.

漢·王充《論衡·氣壽》[12]

A lifetime of a hundred years is a proper length of time for human beings. Just as things will die in autumn, so also there is the proper lifetime of a thing (*wuming*).

Wang Chong, *Lunheng: Qishou* (Han Dynasty)

Tzu Chi teachings build upon the ancient Chinese association of *wuming* with the lifespans of living things by discussing the importance of stretching the lifespans of inanimate things. Thus, Tzu Chi discourse acts to personify or anthropomorphize inanimate objects, reanimating the vast number of nonliving materials and objects that are produced by human industrial machines. Cheng Yen explains in one of her morning sermons,

The Great Enlightened One, the Buddha, what does he want when he comes to the world to educate us? All beings in the world have a life. Let us not destroy lives. Let us not only be vegetarians! Everything has its own life, so how can we cherish it? How can we resurrect each and every thing? These things are like us humans. These lives have pre-existing lives. The lives of living beings go on without end. The fates of objects (物) are similar: they also have causes and conditions. What kinds of persons possess an object? The object will turn into waste if someone throws it away. [Instead, we] can reanimate the object and let it live again, when we pick up the waste object and recycle it.[13]

In her speech, Cheng Yen suggests inanimate things are like sentient beings that have past and future lives, lives that can be prolonged through human beings' intentional actions. Cheng Yen proposes treasuring the lives of inanimate beings with the possibility of reanimating and lengthening their lives (*xu wuming*, 續物命). As for how one is to put the concept into practice and reanimating these objects, Tzu Chi members adopt English terms to talk about the "Five R's": Reduce (少用), Refuse (拒用), Reuse (再用), Recycle (回收), and Repair (修理再用). These concepts represent the new relationship proposed for Buddhist beings and the material, inorganic world. They spiritually reposition modern human beings living in human-made environments and industrial societies.

During the COVID-19 pandemic of 2020, Master Cheng Yen repeatedly uttered this sentence, "This [epidemic] is caused by the actions of human hearts. We should quickly correct ourselves and return to 'We must respect Heaven and Earth; We must love the lives of inanimate beings' (敬天地, 愛物命). I think it will be of great help to this catastrophe." Here, she is again suggesting that the pandemic is a new kind of natural disaster that is divine retribution and collective *karma* for humanity's environmentally destructive behavior. Humanity can undo this karmic retribution of COVID-19 by loving the earth and extending the lives of all the human-made materials we have manufactured, which litter the earth. At the same time, for Tzu Chi followers, the earth as a whole is also made into a living thing that feels pain and suffering, thus deserving of Buddhist compassion. This animation of earth can be seen in Cheng Yen's famous quotation: "We walk on the ground gently because we are afraid the Earth can feel the pain." The sentence provides another example of Tzu Chi Buddhism's unique emphasis on the anthropomorphism of nonliving objects.

Buddhist Ecotechnology

In this section, we will extend our discussion of the Buddhist response to a society at risk by examining the organizational expansion of Tzu Chi Foundation into the area of technology development. In recent years, Tzu Chi volunteers have increasingly promoted environmental solutions through technology. Ecotechnology is a process of technological evolution in which the "[The use of] technology to dominate nature is replaced by the ecological ethic of using technology to harmonize humanity's relationship with nature" (Bookchin 1980: 109). Registered as a Tzu Chi social entity, Da Ai Technology is a nonprofit organization founded in 2008 and manufactures ecofriendly products in Taiwan. The company developed recycled textiles using raw materials from Tzu Chi recycling stations such as polychips and polyester fibers. These textiles are then made into useful things like clothing items and blankets, which are either sold or distributed free of charge in Tzu Chi disaster relief operations around the world. The corporation uses no dyes during the manufacturing process, which lowers energy consumption, carbon emissions, and the use of water. Although there is as yet no Da Ai manufacturing plant in Mainland China, Shanghai volunteers can easily purchase Tzu Chi ecofriendly manufactured products in local Still Thought Bookstores. When members in Shanghai hold the soft eco-textile fleece scarves and blankets, their actions for the green mission of collecting plastic bottles come full circle. Technology provides the final environmental step by demonstrating that a lifestyle of clear consciousness is possible.

In an interview with the manager of Da Ai Technology in Neihu, Taipei, Huang Weishan found that the company is not a profitable business but rather serves as a symbolic representation of "compassion technology." This private enterprise represents an experimental business aiming to fill a niche that the mainstream companies have missed in helping consumers live a lifestyle based in ethical consumption. The number of recycled plastic bottles collected from the Neihu Recycling Station is far below what would be needed for carrying out full-scale manufacturing and production. Da Ai works with over 8,626 Tzu Chi recycling stations in Taiwan. Each year, nearly 2,000 tons of post-consumer PET bottles are collected and recycled by over 200,000 recycling volunteers. Nevertheless, these supplies are still not sufficient. Extra plastic bottles are sometimes purchased to meet the demands of distribution and sales in worldwide Tzu Chi's Still Thought Bookstores. Here, we see that Tzu Chi recycling and eco-technology operations do not share the usual logic of capitalist enterprises, which are based on the profit motive. This non-capitalist business is not anxiously pursuing profit, for it possesses a higher motivating force.

What compensates for the lack of profit is the higher transcendent goal of religio-environmental proselytizing and conversion that eco-technology and recycling make possible. At most Tzu Chi recycling stations in Shanghai, one can find volunteers proudly introducing eco-textile products made from recycled PET bottles from behind their street booths. The volunteers demonstrate to passersby concrete examples of how their tireless plastic recycling efforts have resulted in beautiful polyester scarves. The connections between the labor of waste recycling, Tzu Chi ecotechnology, and remote disaster relief operations financed by eco-technology product sales are always carefully explained to passersby.

IV. STRADDLING THE RELIGIOUS-SECULAR DIVIDE IN SHANGHAI

In his book *A Secular Age*, Charles Taylor (2007) identifies three notions of the modern condition of "secularity." Taylor's first notion, or *Secularity I*, is based on a sociological approach focused on social institutions, where religion becomes separated from the dominant institutions of a society. Jose Casanova's work (1994) was a prominent example of this approach, also called differentiation theory, characterized by the withdrawal of religiosity from the public sphere and the disembedding of religious worldview and rituals from modern economic, political, legal, educational, and family institutions. The second notion, _Secularity II_, describes a decline in personal religious belief and practice, which Taylor likened to the decline of religious involvement and the choice to be religious by many Western youth in the 1960s.

Taylor describes _Secularity III_ as: "a move from a society where belief in God is unchallenged and indeed, unproblematic, to one in which it is understood to be one option among others, and frequently not the easiest to embrace" (Taylor 2007: 3). This third notion represents a shift in the culture away from the assumption that religious faith is the norm, to a condition in which it is possible to not believe in transcendent or supernatural agencies at all, but still be able to live a fulfilling life. This secular condition Taylor calls "the immanent frame," in which people's energies are oriented to worldly concerns, rather than the attaining of lofty transcendent goals that conform with the higher will and intentions of divine authority(ies) or pave the way for the soul in the Afterlife (2007: 542). According to Taylor, Western modernity's secularization process entails the development of the "buffered self," whose self-identity is secure and fixed, and whose ego-boundaries become sharply demarcated and closed to any penetration by spirits, demons, or divine forces. In _Secularity III_, there is also the "rise of a society in which for the first time in history a purely self-sufficient humanism came to be a widely available option . . . accepting no final goals beyond human flourishing" (2007: 18). It is also a social order in which the constitution of society is not based on a metaphysical foundation that is prior to and independent of human actions, nor one that embodies, aligns with, or is subsumed by the cosmic order (2007: 192). Taylor contrasts _Secularity III_, which for him, developed uniquely in the North Atlantic world, where it prevails today, with the situation in most Muslim societies and in India today, where religious faith is the norm, and the majority have not started to question the norm. Indeed, Taylor is wise to acknowledge that his work is based only on examining the history of the European and North Atlantic societies of Christian heritage with which he is familiar.

The experience of secularization in modern China diverges from Taylor's north Atlantic account and instead mirrors many of the findings of Mirjam Künkler and Shylashri Shankar in their edited volume, _A Secular Age beyond the West: Religion, Law and the State in Asia, the Middle East and North Africa_ (2018). This book brought together essays on the modern secularization experiences of non-Western nations around the globe. The editors conclude that the essays highlighted two major historical forces that are generally absent from Taylor's accounting of secularization: (1) the sudden and socially disruptive arrival of external colonizing power(s); and (2) the rise of the modern state, which often accompanies anti-colonial efforts. Due to the anti-colonial push, the modern state often becomes a major institutional actor whose influence extends out to all domains of society. Certainly, these two elements were central in the secularization process in modern China.

There is much scholarship describing the complex processes that transformed modern China into one of the most secular societies in the modern world. In terms of Taylor's *Secularity I*, this process began in 1898 with the late Qing Dynasty imperial edict to seize Buddhist and Daoist temples and monasteries and convert them into modern (secular) schools to build up a strong modern state (Goossaert and Palmer 2011: 43–55). Those temples and ritual sites that were still standing and still used for religious purposes at the Communist Revolution were dismantled or converted to secular uses, especially during Land Reform, the Great Leap Forward, and the Cultural Revolution. Religious rituals and public art and statuary were also expelled from virtually all public spaces. Religious imaginaries and cosmologies, gods, Buddhas, and Daoist immortals, and their rituals were all purged from the primary institutions, such as the state, economic transactions, the educational system, and even traditional Chinese medicine. This means that religious agencies were no longer able to place any checks or moderations on the massive industrial and infrastructural constructions that transformed the natural landscape in the name of strengthening the nation-state and revolution.

As for *Secularity II*, Taylor's accounting renders the decline of religious belief and practices a matter of individual choice, following a larger gradual social trend. In Europe, long before the twentieth century, both Renaissance humanism and the Protestant Reformation promoted cultures of individualism, and in North America, the settling of the frontier led to the ideal of "rugged individualism." Not so in modern China, nor to a lesser extent, in Taiwan. First, the decline of individual belief in China was often part of a conscious and socially engineered process and institutional agency, first by the Guomindang state, and then by the Chinese Communist Party (Yang 2008). The Chinese Communist Party still declares itself "atheist" and forbids Party members from joining any religious organization or professing any religious belief.

Second, unlike post-Reformation Europe and North America, where the culture of the individual developed in strength, religious life in premodern China and in post-Mao rural communities was much more a life force of local villages and communities, lineage and clan organizations, and guilds and occupational groups. The culture of the individual was not introduced into China until the May Fourth Movement (1919–1930s), which was limited to urban educated circles, and was quickly ended with the Communist Revolution. Thus, in modern China, although the "buffered self" did develop

to prevent the easy crossing of boundaries between humans, deities, ancestors, animals, and demons and other divine agencies, the agency of individual
choice is a very recent phenomenon in post-Mao China. It was religious
group commitments and traditional ritualized localisms that were often seen
as obstacles to building up the nationalism that was felt to be needed to construct the modern monolithic state (Yang 2020).[14] Thus, we can say that *Secularity I* and *II* took place in modern China and Taiwan as part of the larger
process of the rise of the modern nation-state, and they were often directed
by the state. However, after the Guomindang moved to Taiwan in 1949, their
secularization efforts were not as severe or systematic as on the Mainland.

Taylor's *Secularity III*, where religious life is an option or choice, could
apply to *urban* China during the first half of the twentieth century. After the
Communist Revolution of 1949, during the Maoist era, *Secularity III* did
not really apply to China, for there was virtually *no* option to believe, so
strong and systematic were the secularization process and the commitment
of the state to secularism, the ideology of the party-state. Chinese society
had moved from a modernizing society of having the option to believe or
not believe before the Communist Revolution, to one where there was only
one option, to *not* believe. Thus, the Maoist era was virtually *the opposite* of
Taylor's *Secularity III*.

Since the 1980s, in the post-Mao era, Secularity III is again relevant in
China, in that there is now the option to believe. However, this option differs
from Taylor's North Atlantic, especially the American scenario, because in
China, it takes place in a context where the norm is overwhelmingly to *not*
believe. Much more than in contemporary Europe, in the extreme secularity
of Chinese urban society today, *not* believing is what is "unchallenged" and
"unproblematic," while believing needs to be justified, and is therefore "frequently not the easiest to embrace."

In Taylor's formulation, *Secularity III* is described as fundamentally
pluralistic, with multiple forms of religiosities and religious-secular hybridi-
ties to choose from. This is true for Taiwan, but not for Mainland China. In
contemporary China, those who choose to believe are restricted to only five
officially recognized religions: Buddhism, Daoism, Catholicism, Protestantism, and Islam. All five religious communities are governed by quasi-state,
quasi-civil society bureaucratic administrations that must comply with
policy directives from Beijing's state and party-run religious bureaucracy.
Those religions that fall outside these officially legitimate religions, whether
Greek Orthodox, Mormon, Bahá'í, or Tzu Chi, must lead an unpredictable
and sometimes furtive life, at times accepted or tolerated, and occasionally
harassed by state authorities. Furthermore, like these other anomalous religions which have no proper state-administered bureaucracy, Tzu Chi also
operates with the disadvantage of being a foreign creed, coming from outside

Mainland China. These anomalous religions introduced from abroad must deal with state surveillance and the suspicion that they harbor unhealthy modes of thought that may contribute to political destabilization in China. Next, we will explore how Tzu Chi's religious environmentalism negotiates with the specific conditions of *Secularity III* in urban Shanghai.

Tzu Chi and *Secularity III* in Shanghai

Today, while both the Chinese state and Chinese urban society remain staunchly secular, there are also modest developments in religious revival. In the post-Mao era beginning in the 1980s, rural migrants who have moved to Shanghai in search of jobs and an urban life have reintroduced various religious rituals and faiths to secular Chinese cities. China's opening to the global economy has also allowed religious forces from the outside world to enter, such as various forms of Christianity from North America, Europe, and South Korea, and Buddhism from Taiwan and Southeast Asia. At the same time, state and Party regulations and policies allow for the "freedom of religion," which was enshrined in the original 1954 Constitution, but was barely put into practice during the Maoist era. Thus, in Shanghai today, we find a new mixture of religious cultures that have altered the state-approved religious landscape and intensified existing tensions between the state and some religious groups.

What is unique about urban Chinese contexts, as compared with the North Atlantic or other Asian societies, is the severe state secularization of the Maoist years. This means that for recent local Tzu Chi converts in Shanghai, about 80–90 percent of them are first-generation religious followers. They were not born into Buddhist families or local communities, nor did they have any previous faith affiliation. It also means that since these converts live in an urban environment that is so predominantly secular, they can easily slide back into a secular self, or swing back and forth between a religious or secular life, values, and commitments.

Even before arriving in Shanghai in the early 1990s, the Tzu Chi Foundation was already the most secularized of Taiwan's major Buddhist organizations. It's focus on spiritual cultivation through humanitarian action in the secular world was a radical concept when Master Cheng Yen first expounded it in 1966. There is a direct Buddhist lineage descent line from her to the great Buddhist monk reformer, Master Taixu (太虛 1890–1947) in China. Master Cheng Yen's own Buddhist teacher and patron was Master Yin Shun (印順 1906–2005), who, in turn, was a disciple of Master Taixu when they were both living in Mainland China. Taixu espoused radical Buddhist monastic reforms from 1912 to the 1940s, advocating what has come to be known as "humanistic Buddhism" (*renjian fojiao* 人間佛教). In light of what Taylor wrote about secularization in the modern West as introducing a new order of "exclusive humanism," in which human flourishing becomes for the first time the only concern (Taylor 2007: 18–21), we suggest a different translation for Taixu's reform of Buddhism. The usual translation of "humanism" in modern Chinese is "human ethics-ism" (*rendao zhuyi* 人道主義). The Chinese words *renjian* (人間), while invoking the human, also emphasizes the *place* where humans reside, amidst other humans, that is, in this temporal human world. Therefore, we propose a better translation would be "humanistic this-worldly Buddhism" to emphasize this temporal life on this planet earth, rather than on humanism, which is more important in the secularization of the Christian West. In most forms of contemporary Chinese Buddhism, no matter how secularized, there is still the adherence of the non-humanistic Buddhist diet of vegetarianism and the invoking of "sentient beings," which relativize the position of humans among other life forms.

Taixu's ideas in the early twentieth century were too early for their time, and they were not accepted, let alone implemented, by the majority of Buddhist clerics or laity during his lifetime (Pittman 2001; Ting 2007; Ji 2013; Laliberte 2015). However, in the new environment of Taiwan in the 1960s, the secularizing efforts spearheaded by Master Cheng Yen of Tzu Chi and other Buddhist leaders such as Master Hsing Yun of Fo Guangshan, another globalizing Taiwanese Buddhist organization, fell on fertile soil, and have developed since. The model of religiosity offered by Tzu Chi departed from traditional Buddhism in downplaying the afterlife, Buddhist miracles, scripture chanting, rituals, the cults of different Buddhas and Bodhisattvas, and the role of Buddhist clergy. In the 1990s, many of Tzu Chi's environmental activities, such as street cleaning and sanitation, were generally considered unorthodox endeavors for Buddhist communities. In the name of environmentalism, Tzu Chi's teaching already prohibited the use of incense and burning of spirit money when making religious offerings. Tzu Chi's shift toward religio-enviornmentalism is also apparent in other traditional practices. The lunar festival of the fifteenth day of the seventh month, traditionally known as the Buddhist Yulan Basin Festival (盂蘭盆法會) or popular religion's Ghost Festival (鬼節), was renamed by Master Cheng Yen to the "Auspicious Month to Commemorate Ancestors." Instead of burning incense and spirit money, Tzu Chi promotes the giving of alternative offerings, such as charitable deeds, protecting animals, and joining the plastic-free movement.

Already in Taiwan, Tzu Chi was a predominantly lay organization whose members were this-worldly, practicing a form of engaged Buddhism that encouraged direct social activism and volunteer humanitarian work to relieve suffering in the world. Tzu Chi called upon its clerical members to "reject lay offerings [of wealth and food], and discard [economic reliance on] ritual performance" (不接受供養，不幹經懺). Although the teachings of Tzu Chi

Foundation do not publicly criticize "traditional" Han Buddhism, its practice of lay-centered social services has directly challenged the ideological supremacy of Mahāyāna Buddhist temples and the leadership role of Buddhist clergy in Taiwan, as well as in China. The predominant role of lay activists in the Tzu Chi organization may have facilitated its movement from Taiwan to Mainland China, since China's state religious bureaucracy that oversees all Buddhist temples and monasteries across the country probably felt less threatened by lay Buddhists than an overseas institutional clerical authority.

In its drive to accommodate itself to a secular society, Tzu Chi Foundation also extended and reinterpreted two important Buddhist concepts: "wisdom" (*zhihui* 智慧 *prajñā* in Sanskrit) and "self-cultivation" (*xiu* 修). In Mahayana philosophy, "wisdom" is the understanding of the true nature of the phenomenal world: that it is impermanent; that it engenders much suffering from the desires that it arouses; and that it is ultimately "empty," therefore, people should avoid getting attached to it and the desires it elicits. Tzu Chi substitutes this abstract esoteric Buddhist discourse with the more pragmatic and relatable ideas of "wisdom and clarity" (慧明) and "wisdom about life" (慧命) that appeal to ordinary people grappling with modern life and the risk society. It links up the "cultivation of wisdom" (修慧明) with the problem of how to "nurture a Buddhist heart." According to Master Cheng Yen,

見苦知福，修慧明諦，人間菩薩不只是能知福，還能修智慧，智慧要
從真理而來，而真理都是在人心。不知道如何啟開內心真諦，就要往
外見苦知福，往外修慧明理.

Whether one experiences suffering or prosperity, one still needs to cultivate wisdom. Bodhisattvas living in this world must not only experience happiness, they must also cultivate their wisdom. Wisdom can only come from true experiences, and truth comes from the human heart. When one does not know how to open the truthful heart, then one must go outside [oneself] to see suffering and know happiness, go outside to cultivate wisdom.[15]

Here, Master Cheng Yen has reinterpreted "wisdom" from its earlier Buddhist association with a retreat from the illusory world toward an inner contemplative life of meditation and scripture study in monastic spaces, into an exhortation to engage with the outside phenomenal world through volunteer work. In Master Cheng's discourse, the notion of "self-cultivation" (修) has also undergone transformation from the act of meditation and scripture study to active forms of social service and charity as a public expression of religious cultivation. Among Tzu Chi's forms of self-cultivation is service in environmental protection. She often emphasizes the relationship between reviving decency, virtue, and environmental protection, using modern global slogans in her speeches, such as "Love the Earth" (愛地球). For Tzu Chi

Buddhists, this phrase holds a double meaning. It encompasses the secular environmentalist understanding, while also resonating with the Buddhist notion of "cultivating wisdom" (修慧明).

The Tzu Chi emphasis on humanitarian action in the world is encapsulated in Tzu Chi slogans that Huang Weishan heard uttered frequently by Tzu Chi volunteers in Shanghai when she asked why they were volunteering: "When things are right to do, just do it, and it's right!" (對的事，做就對了!). Tzu Chi volunteers also frequently used this phrase: "Learning by doing, learning gives rise to Awakening" (做中學，學中悟). Here, the Buddhist notion of gaining Enlightenment and attaining Buddhahood is shifted from the older emphasis on detachment, contemplation, and emptying out the mind, to direct engagement with the social and natural world through action. Tzu Chi Commissioners also used these phrases to give members the courage to follow through on their ideals. Finally, there was also the sense that members do not always need to comprehend everything before they act. Huang also found that Commissioners invoked these sayings as ways to encourage Tzu Chi members to follow the master's instructions without too many questions, as revealed in this phrase, "Follow closely the footsteps of the Superior One; When things are right to do, just do it, and it's right!" (緊跟上人的腳步，對的事情，做就對了!).

From the movement's early days in Taiwan, Tzu Chi followers were required to serve as volunteers if they intended to convert. This practice has been followed in the Tzu Chi tradition ever since. Therefore, the identity of Tzu Chi followers can often be less of a devoted Buddhist, and take the form of a more secular identity, a Tzu Chi member belonging to a community of humanitarian volunteers. The teachings of Tzu Chi encourage one to act on the problems of the secular world instead of avoiding them. However, the secular volunteer efforts of helping the needy have the purpose of "cultivating one's spirituality." In 2018, Huang Weishan held an interview in Shanghai with a high-profile Tzu Chi Commissioner, Sister Chiu. Her explanation of the relationship between Buddhist cultivation, charitable action, and the Tzu Chi organization elucidates how the ritualization of secular actions leads back to a religious life.

> The Tzu Chi Foundation is more of a spiritual cultivation group, not simply a humanitarian organization. It is Master Cheng Yen's hope that our practitioners will cultivate themselves through benevolent works. There are so many schools of Buddhism nowadays. She has established a separate and different Buddhist denomination called the Tzu Chi School (慈濟宗), where the method of cultivation is the Still Thought Lineage (靜思法脈). In this Lineage, we focus on the teachings of the *Sutra of Immeasurable Meanings* 《無量意經》 and the *Lotus Sutra* 《妙法蓮華經》. For Tzu Chi people, the true way to following a Bodhisattva path is to "walk into society" (走人群). Why do we repeatedly utter the words "faith, vow, and practice" (*xin-yuan-xing* 信，願，行)? First of

all, you must believe in the method [of the Tzu Chi School]. The foundation of this school is that you must take a vow. You will engage in cultivation "among people," not by chanting sutras or sitting in meditation in a peaceful environment under a tree. "Faith" (*xin*): first of all, this means that you must believe in this School. *Yuan* means that you must take a vow. "Practice" (*xing*) means that you must engage in cultivation by going among the people to walk on a Bodhisattva path.

Sister Chiu's statement shows how the Tzu Chi Foundation strongly encourages people to absorb scriptural teachings by acting upon the problems of the secular world. There is a new emphasis on "practice" or "doing," but these still revert back to the scriptural teachings and religious faith.

The Tzu Chi Foundation is still in the process of learning to integrate secular principles into its mission and language, while balancing it with ancient Buddhist religious teachings. The organization adopted secular activists' language after it was granted a consultation status with United Nations in 2010 and accreditation as an observer of the UN Environment Assembly (UNEA) in 2019. During its interaction with international governmental and nongovernmental organizations, Tzu Chi generally adopts a secular language. Yet, as we have shown earlier, at its heart, among its core dedicated lay volunteers, Faith Corps, and Commissioners, it remains a religion-based humanitarian movement with some continuing other-worldly urges.

V. CONCLUSION

In this chapter, we demonstrated how the Tzu Chi Foundation is a manifestation of a new kind of Buddhism environmental ethics, operating through the spiritual cultivation of the self, in the private domestic sphere of the home and kitchen, in the public spaces of residential communities, urban streets, and certain private workplaces in Shanghai. Tzu Chi rank-and-file members create meaning through daily behaviors, such as by adopting ethical eating, ethical consumer habits of consumption, lifestyle changes, recycling, volunteer actions, and even non-profitable eco-technology business ventures. We showed how Tzu Chi has adapted and adjusted ancient Chinese Buddhist teachings and scriptures to formulating a new religio-environmental ontology for our rapidly depleting natural environment and increasing dangers of global climate change. This alternative ontology builds upon ancient Buddhism's abundant environmental potentials, such as the incorporation of all sentient beings into the realm of "Buddha nature," doctrines of the impermanence and illusoriness of the phenomenal world, and the need for "self-cultivation," which is translated as exerting real, direct humanitarian and environmental action in this world and this life. Tzu Chi ontological adaptations also include the reinterpretation of the Buddhist notion of "collective *karma*" to explain the causes of natural disasters and environmental degradations leading to hunger, disease, and war. An impressive ontological innovation unique to Tzu Chi is the move to "animate" physical synthetic objects, the results of the massive pollution of human industrial manufacturing. In encouraging people to ascribe life force to inanimate waste objects (*wuming*), imbuing them with sentience and the ability to be transformed into new lives, Tzu Chi both justifies its recycling projects and prolongs the time before they may be buried under the earth or thrown into the oceans and rivers.

Finally, on the process of secularization in Chinese modernity, we addressed how it differs in the *Secularity I* and *II* modes that Charles Taylor laid out for the North Atlantic modern West. The post-Mao resurgence of religiosity in urban China, such as in contemporary Shanghai, falls within Taylor's mode of Secularity III, an overall secular context wherein certain people and groups choose to be religious. Tzu Chi teachings and activities demonstrate the inroads of secularization upon traditional Buddhism: they have reinterpreted spiritual cultivation as action in the world (not contemplation), and they propel believers away from a focus on past and future lives, toward attaining Buddhahood, or at least making a difference, in *this* life. However, in contrast to Taylor's Christian cultural heritage of the North Atlantic, Tzu Chi's Buddhist religio-environmental teachings resist the "exclusive humanism," that is, a core feature of modern Western secularity.

While the Tzu Chi movement represents a reformed and secularized this-worldly modern Buddhism, we also revealed that the religious transcendent motivations continue to be important in Tzu Chi actions. The push to protect the environment and save planet earth is not fueled solely by a concern for biological existence, but more importantly, because earth is understood as the one place in the universe where our rebirth as a human being provides us with the rare opportunity to cultivate ourselves toward attaining "wisdom" and Enlightenment, that is, transcendence. We have shown how Tzu Chi Foundation's straddling of the religious-secular divide confirms Taylor's assertion that the modern shift into an ontology of the "immanent frame" does not close *Secularity III* off from any transcendental claims or imaginaries (2007: 550). The Tzu Chi Foundation, its dedicated clerical managers and lay volunteers, provide a living testament to the continued viability, perhaps even desirability, for transcendence in environmental efforts in a secular world.

NOTES

* Aka Weishan Huang: http://orcid.org/0000-0003-3804-9523.
1. The Buddhist Tzu Chi Foundation is the organization's official name in Taiwan. Tzu Chi Foundation is the name of the organization in China, where it is registered as a non-governmental organization in 2008.

2. See Julia S. Huang's ethnographic study of the emotional relationship between the charismatic Master Cheng Yen and her Tzu Chi devotees (Huang, J. 2009) and An Pham's PhD dissertation (directed by Mayfair Yang), based on fieldwork in Taiwan with Tzu Chi's Ta Ai Television Station, from both the media production and reception angles (Pham 2017).

3. "2012 Global Tzu Chi Geographic Locations of Branch and Chapter" May 2013. https://www.tzuchi.org.tw/en/index.php?option=com_content&view=article &id=1092&Itemid=282&lang=en (Accessed Oct. 19, 2020).

4. Commissioners are lay leaders who have been certified. Faith Corps (*cicheng* 慈誠) are formed by male commissioners. Both Commissioners and Faith Corps members are registered lay volunteers who, after two years of apprenticeship, become fundraisers for the organization and leaders in various divisions of the organization's work in the field. Their training includes participating in Tzu Chi missions, spiritual seminars, Buddhist etiquette, and studying the Ten Precepts of Tzu Chi. They also engage in visiting the poor, and fund-raising.

5. Zi-hao Ye, "A History of Environmental Action" in *Tzu Chi Culture & Communication Foundation.* https://web.tzuchiculture.org.tw/?book=&mptce=8383 (Accessed Nov. 11, 2020).

6. The regulation was called the "Shanghai Municipality Regulations on Life and Management of Garbage" 《上海市生活垃圾管理條例》. Although it was passed on January 31, 2019, but was not implemented until later in the year.

7. The Shanghai municipal regulations regarding domestic solid wastes required that all households sort their garbage into four categories: (1) Recyclables: wastes such as waste papers, plastics, glass products, scrap metals, and fabrics, which are suitable for recycling and cyclic utilization; (2) Hazardous Waste: discarded batteries, lighting bulbs, drugs, and paints with their containers, which can cause harm to human health or the natural environment; (3) Household Food Waste, Perishables, or Organic Waste: raw food materials, leftovers, expired foods, melon kernels, garden waste, and Chinese medicine, etc.; and (4) Residual Waste, or Other: https://news.ch ina.com/zw/news/13000776/20200612/38343596.html (Accessed Nov. 18, 2020).

8. See http://www.tzuchi.org.tw/epa/index.php/2016-08-11-02-39-6/1653-2020 -01-30-03-28-22 (Accessed Nov. 17, 2020).

9. Hsiao, Chiuling 蕭秋玲著. "四通八達" ("Four Crossings and Eight Arrivals") in 《慈濟傳播人文志業基金會》 (*Tzu Chi Culture & Communication Foundation*). March 2018. https://web.tzuchiculture.org.tw/?book=9040&mmtao=7224# .X6xbAchKiUk (Accessed Nov. 11, 2020).

10. *Ibid.*

11. *Ibid.*

12. In "Chinese Text Project" （中國哲學書電子化計劃）https://ctext.org/lun-heng/qi-shou/zh. (Accessed on Sep 13, 2020).

13. "Master Cheng Yen's Morning Sermon for Volunteers" Online text published by Tzu Chi Foundation. Nov. 14, 2019, https://www.tzuchi.org.tw/證嚴上人/法音宣流/志工早會開示/item/22921-不輕己靈-惜緣度眾 (Accessed Sep 13, 2020.)

14. Thus, in Chinese modernity, the energies and labor of the masses had to be diverted away from other-worldly and transcendent concerns, and rechanneled into a focus on this temporal world, and loyalty to the imperiled nation.

15. https://www.tzuchi.org.tw/%E8%AD%89%E5%9A%B4%E4%B8%8A%E
4%BA%BA/%E6%B3%95%E9%9F%B3%E5%AE%A3%E6%B5%81/%E5%BF
%97%E5%B7%A5%E6%97%A9%E6%9C%83%E9%96%8B%E7%A4%BA/item/
1651-%E4%BF%AE%E6%85%A7%E6%98%8E%E8%AB%A6-%E8%BA%AB
%E5%BF%83%E7%9A%86%E6%B5%B4%E4%BD%9B (Accessed Nov. 2, 2020).

BIBLIOGRAPHY

Beck, Ulrich. *What Is Globalization*. Polity, 2000.
———. *The Metamorphosis of the World*. Polity, 2016.
Bookchin, Murray. *Toward Ecological Society*. Black Rose Books, 1980.
Casanova, José. *Public Religions in the Modern World*. Chicago: University of Chicago Press. 1994.
Ho, Peter. *Leapfrogging Development in Emerging Asia: Caught Between Greening and Pollution*. Nova Publishers, 2008.
Ho, Peter and Richard Edmonds. *China's Embedded Activism: Opportunities and Constraints of a Social Movement*. New York: Routledge, 2007.
Huang, Julia C. 2009. *Charisma and Compassion: Cheng Yen and the Buddhist Tzu Chi Movement*. Cambridge: Harvard University Press.
———. "Buddhism and its Trust Networks between Taiwan, Malaysia, and the United States." *The Eastern Buddhist* 44, no. 2 (2013): 59–76.
Huang, Weishan. "The Place of Socially Engaged Buddhism in China - the Emerging Religious Identity in the Local Community of Urban Shanghai." *Journal of Buddhist Ethics*, no. 25 (2018): 531–568.
Ji, Zhe. "Zhao Puchu and His Renjian Buddhism" in *Eastern Buddhism*. December, 2013.
Künkler, Mirjam and Shylashri Shankar. "Introduction." In *A Secular Age Beyond the West. Religion, Law and the State in Asia, the Middle East and North Africa*, edited by Mirjam Künkler, John Madeley and Shylashri Shankar. Cambridge University Press, 2018.
Laliberté, André. " 'Buddhism for the Human Realm' and Taiwanese Democracy." In *Religious Organizations and Democracy in Contemporary Asia*, edited by Tung-ren Cheng and Deborah Brown. Routledge, 2015.
Pham, An Quoc. *Buddhist Television in Taiwan: Adopting Modern Mass Media Technologies for Dharma Propagation*. PhD Dissertation in East Asian Languages and Cultural Studies, filed at University of California, Santa Barbara, 2017. https://www.escholarship.org/uc/item/57c3j5xx.
Putnam, Robert D. "Bowling Alone: America's Declining Social Capital." *Journal of Democracy* 6, no. 1 (1995): 65–78.
Putnam, Robert D., Robert Leonardi, and Raffaella Y. Nanetti. *Making Democracy Work: Civic Traditions in Modern Italy*. Princeton, 1993.
Taylor, Charles. *A Secular Age*. Harvard University Press. 2007.
Ting, Ren-Chieg. "Renjian Buddhism and its Successors: Toward a Sociological Analysis of Buddhist Awakening in Contemporary Taiwan." In *Development and*

Buddhist Environmentalism and Civic Engagement   99

Practice of Humanitarian Buddhism. Mutsu Hsu and Jinhua Chen, edited by Hualien, pp. 229–268. Taiwan: Tzu-Chi University Publisher, 2007.

Weller, Robert P. *Alternate Civilities: Democracy and Culture in China and Taiwan*. Boulder, CO: Westview Press, 1999.

Yang, Mayfair. "Introduction." In *Chinese Religiosities: Afflictions of Modernity and State Formation*, edited by Mayfair Yang. Berkeley: University of California Press, 2008.

Yang, Mayfair. *Re-enchanting Modernity: Ritual Economy and Society in Wenzhou, China*. Duke University Press, 2020.

II

SACRED SITES AND *FENGSHUI* LANDSCAPES

Chapter 3

Fengshui and Sustainability
Debating Livelihoods in the Qing Dynasty
Tristan G. Brown

I. INTRODUCTION

In late imperial China, how and why did people oppose mining? How did
the state respond? Looking at a range of memorials and legal cases preserved
in archives in the People's Republic of China and Taiwan concerning mining disputes, this chapter examines notions of sustainability in late imperial
China, particularly in relation to debates over what Qing officials called the
"livelihoods of the people" (*minsheng* 民生). This chapter demonstrates that
both proponents for mining and opponents to mining cited the well-being of
people and communities to express policy preferences in official settings.
During these discussions, *fengshui* was regularly invoked by opponents
of mining, both in the imperial bureaucracy and in local society. The state
responded seriously to these memorials and petitions. Building on this observation, the chapter explores the range of debates over *fengshui* and mining
in the late imperial period to challenge the narrative that China's history has
been marked by linear environmental degradation or a unilateral advocacy for
the domination of nature. In fact, ethical, moral, and political issues related to
sustainability were debated and sometimes enforced through the legal system.

Official rhetoric concerning the "livelihoods of the people" – an ancient
concept in political thought – was ubiquitous in the Qing bureaucracy of
the eighteenth and nineteenth centuries.[1] Loosely defined as the material
conditions of life related to the standard of living, there was a widespread
sense, particularly during and after the reign of the Qianlong Emperor (r.
1735–1796), that *minsheng* was threatened in many regions due to the mount-
ing pressures of population growth on scarce resources. During Qianlong's
reign, the statesman Hong Liangji (洪亮吉) (1746–1809), sometimes compared to the English economist Thomas Malthus (1766–1834), composed a

well-known essay in 1793 on the social issues surrounding the limitations of resources around the time the empire's population crossed 300 million people. In it, Hong wrote that "flooding, drought, disease, and epidemics, these are the ways in which Heaven and Earth temper the problem [of over-population]" [水旱疾疫，即天地調劑之法也] (*Zhongguo jingji sixiangshi ziliao xuanji* 1990: 502). Hong warned that if the imperial state did not meet the growing material demands of the people, heaven would itself interfere to remedy the problem. The era in which Hong was writing saw an elevation of concerns over the livelihoods of the people in light of the dramatic demographic shifts that marked that time.

Hong Liangji was not alone in expressing concerns over the peoples' livelihoods. As William Rowe has discussed for the eighteenth-century Qing stateman Chen Hongmou (陳宏謀) (1696–1771), officials such as Chen sought to support the livelihoods of the people by providing "the most conducive possible environment for the people's productive capacity" (Rowe 2001: 189). Avoiding onerous taxation, cautiously permitting the opening of private mines, and ensuring stable employment for laborers were all components of the official pursuit for securing the peoples' livelihoods.

But other ways to protect the people's livelihoods existed as well. One discourse frequently invoked by officials and commoners alike was cosmological in nature. *Fengshui* ("wind and water"), often translated as Chinese geomancy, had been practiced for centuries to orient houses and graves in relation to the surrounding environment for the maximal benefit of humans. By the Qing period, it was often cited during discussions and disputes over mining in order to maintain the proper balance between the cosmic forces within a landscape. Qing officials were no strangers to this discourse. In 1762, for instance, Chen Hongmou, a routine advocate for opening mines to support of the people's livelihoods, agreed with a county magistrate's suggestion to ban mining around Hunan's Jiuyi Mountain in order to protect its dragon vein—a cosmological channel thought to connect mountain and river systems. Part of the magistrate's logic for protecting the dragon vein involved securing uncontaminated water sources around the mountain for local irrigation (Lin 2014: 226).

The image that emerges from the Qing archival record is one of regular debate and compromise. Presumably nearly every Qing official at least professed concern about securing the peoples' livelihoods during their tenures. The trenchant question during the Qing was in how to best do so. Advocates of mining in the palace bureaucracy argued that the opening of new mines would employ landless laborers and generally improve economic conditions. The detractors of such proposals countered that they, too, were concerned for the livelihoods of the people in their advocacy for the protection of *fengshui*, dragon veins, and ancient tombs. To compromise these sites was to potentially endanger the fortune and well-being of local communities and undermine the broader imperial order by ignoring powerful natural forces (Perdue 2010: 101–119). As mentioned earlier, Qing officials such as Chen Hongmou held highly nuanced views on many of these issues, and stances could vary depending on the particular case.

To be clear, advocates for mining during the Qing were seldom arguing for unchecked development. Likewise, the detractors of mining proposals were not advocating for anything like environmental protection or the maintenance of wilderness. They were both rhetorically appealing to the same imperial ideal of securing the people's livelihoods through the pursuit of imperial-era notions of productive sustainability and community stability. When officials advocated for the opening of private mines, they were often seeking to make the labor market of a particular region sustainable to prevent social unrest. When they agreed to ban mining on a particular site in order to protect *fengshui*, they sought to protect the sanctity of important sites, to maintain the balance of resources along a natural landscape for the agricultural productivity of future generations, and to assure locals that their good fortune was protected by the state.

Actions deemed sustainable in one era may not be sustainable in the next, and questions concerning the durability of commercial practices involving natural resources in the imperial-era are receiving significant scholarly attention (Miller 2020; Zhang 2021). With these recent works in mind, this chapter employs the concept in a relative sense, both in relation to alternative discourses available at the time and ones prevailing at present. Discussions about the ecological consequences of human development today are often couched in a binary wherein economic profit is pitted against environmental damage. We must begin by recognizing that this binary does not perfectly apply to China historically. In the case of objections to mining during the Qing, perceived social gain was always pitted against a perceived social cost. It was up to the officials of the imperial state, and sometimes the emperor, to negotiate the ethical applications of these competing claims, with ethical simply meaning that the state would typically frame its decision, regardless of what it was, in moral terms while concealing the political contingencies that had led to a verdict.

Dipesh Chakrabarty has suggested that climate change spells the "collapse of the Age-old Humanist Distinction between Natural History and Human History" (Chakrabarty 2009: 197–222). It is true that this distinction applies well to the binary between humanist and scientific knowledge that underpinned the Enlightenment project. But it is worth considering that history as traditionally narrated in China regularly made note of environmental forces, whether in official dynastic histories or popular local ones. Just as the severing of dragon veins was thought to potentially end dynasties, the opening of mines could potentially summon grave misfortune for their surrounding communities, manifested through plague, illness, and crop failure. Such tales indeed were woven into popular histories of dynasties and the contents of family genealogies. The felling of an ancient *fengshui* tree in front of the tomb of the first Ming emperor in 1642 was popularly thought to have summoned the collapse of the dynasty two years later (Brook 2010: 133). This dramatic narrative of the dynastic transition reflects the landscape that officials negotiated as they processed mining requests or litigated mining lawsuits.

This chapter is divided into three sections, each exploring a different facet of geomantic discourse as it applied to peoples' livelihoods and sustainability. The first examines the perceived inauspicious effects of poor *fengshui* on human communities, detailing how the land was invoked as a living body. The second section introduces Qing imperial edicts related to mining and *fengshui*, showing imperial-era debates over mining in relation to the livelihoods of the people. The final section concludes with some county-level legal cases, highlighting the invocation of these principles in local society.

II. HEALTH AND LAND IN CHINESE HISTORY

The links between graves, *fengshui*, and human health can be traced back well in imperial history. To provide just one example, the Song dynasty treatise, *Complete Good Prescriptions for Women* (婦人大全良方), published by Chen Ziming (陳自明) in 1237, provides three possible reasons for the failure to conceive sons:

> The first is, if the grave is not sacrificed to; the second is, if the fate of the husband or wife subdues the other; the third is, if the husband is struck with an illness or the wife contracts measles. These factors will all produce no sons (Chen 2007: 190).

一者，墳墓不嗣[祀]；二者，夫婦年命相剋；三者，夫病婦疹，皆令無子。

Embedded in this passage is a fundamental principle of Chinese cosmology: the well-being of graves was seen as intimately linked with the health of the living. It is well-worth considering that the aforementioned passage, which quotes a medieval text, did not derive from a geomantic manual, but rather a medical treatise. Infertility in women, commonly seen as the product of an "evil spirit" (*yinxie* 陰邪) formed by an excess of *yin* and a lack of *yang* energy, necessitated an examination of the energy balance (the *fengshui*) inherent in the residence of the dead—the *yinzhai* (陰宅; the family grave). This brief excerpt helps us reconstruct the logic that officials confronted while processing lawsuits about harmed earth veins thought to endanger the health of living people.

Medical and geological concepts were referenced on a continuum under the physiological-geographical concept of meridians or veins (*mai* 脈), which were thought to channel *qi* throughout the human body and through the soils that constituted the natural environment. Geomantic veins were identified widely across rural landscapes around graves, ancestral halls, and temples. These veins typically denoted a place where the natural environment (a river, tree, etc.) was deemed to hold a close connection to the built environment (a grave, temple, etc.). *Fengshui* was invoked to analyze the balance of *qi* in a landscape by ascribing veins and built structures with compass directions based on the ten heavenly stems (*tiangan* 天干) and the twelve earthly branches (*dizhi* 地支) to orient graves and houses for the benefit of people. The severing of these veins could produce serious medical symptoms and the sudden loss of wealth. As we will see later in this chapter, such misfortunes were routinely cited as evidence of wrongdoing in legal disputes.

A geomantic manuscript from Qing Sichuan details some of the outcomes of poor fengshui or the severing of an earth vein. A wide range of medical conditions could be traced back to the *fengshui* of a gravesite:

> If the *mao* [direction] is too much, then the men in the household will cough up blood. If the *yi* [direction] is too much, then women in the household will cough up blood. If *yi* is in excess, then you will have many daughters, if *mao* is in excess, then you will have many sons . . . . If you see that two waters come from *yi* and *chen*, you can judge that the household will produce a person who is crippled and a person who stutters. . . . If the Bright Hall [of the grave] is in the *yin* and *jia* directions, [the family] will be afflicted with leprosy, the death of the mother in childbirth, death from hanging, or death from a stroke. (*Dili yuanxu* 地理原序, §3.2–§4.1)

> 卯多男人吐血；乙多婦人吐血。乙多生女多；卯多生男多...
> 見有乙辰二水朝來，　斷他出拐腳人，出結舌人...若有明堂在寅甲方，
> 出麻瘋, 生產死, 吊頸死, 中瘋死。

As seen in the aforementioned description of the potentially inauspicious geomantic orientations, the grave was thought to influence human health for a range of people within a household. Specific geomantic formulas were thought to hold particular significance for men and women respectively, and here again childbirth was invoked in relation to the family grave.

Beyond medical issues, a grave's poor *fengshui* was also thought to be connected to various failures in the agricultural system, be it through irrigation, drought, flooding, or the sudden loss of landholdings. Alternatively, good grave *fengshui* had the capacity to secure substantial wealth:

> When the *chen* dragon and the *xu* water spread through the farming estates, [you will be] wealthy to the extent that it approaches the wealth of the country and possess much taxable grain. If the resplendence of the *xu* water and the *xu* hill illuminates the [grave's] cavern, a scholar of the Hanlin Academy will approach the splendor of the Dragon. If the *qian* dragon encounters the beauty of this hill and water, both successful examination candidates as well as (male) descendants and wealth will fill the entire township. (*Dili yuanxu* 地理原序, §15.1)

> 辰龍戌水廣田莊，富堪敵國多稅糧。戌水戌砂光照穴，翰林學士近龍光。乾龍遇此砂水秀，科甲丁財遍滿鄉。

From the aforementioned texts, we can observe that *fengshui* manuals stressed that the gravesite was closely associated with physical and financial well-being—the same things that the people's livelihoods ideally secured. Because the grave was seen as sustaining the well-being of a family, it was incumbent upon families to maintain the well-being of the grave by carefully protecting the environmental characteristics that defined its initial creation.

This particular point is key for understanding chains of events that *fengshui* rationalized in imperial Chinese society. While there were debates over particular landscapes and over the abilities of ritual specialists, there was a general consensus between medical and geomantic texts that disease could arise from the environmental characteristics of gravesites, houses, and village settlements. We might recall that "flooding, drought, disease, and epidemics" were identified by Hong Liangji as the ways that heaven and earth tempered the problem of overpopulation. While people today might be inclined to assume that the obvious answer to Hong's identified problem would be to increase agricultural and mineral production, in Qing times, that solution was not the only obvious remedy. Protecting the cosmological balance of a landscape to prevent floods, droughts, and diseases from occurring was also a compelling prescription well into the nineteenth century.

In sum, landscapes in imperial China were cosmologically embedded with cosmological significance through the language of *fengshui*. The spread and transformations of this cosmology did not lead to a romanticizing of nature or demands to protect the environment in its entirety; to the contrary, as seen in the ritual texts and historical records given earlier, the natural and built environment constantly needed to be tamed, sustained, and managed, lest dangerous afflictions or unfortunate events arise. Like the human body, elements of the landscape were subject to sudden or gradual illness and death, and proper geomantic diagnosis, as in medicine, depended upon the correct analysis of the relevant veins. Archival records reveal at least four possible ways for these earthly veins to be compromise: improperly erecting a building or a grave, cutting down a tree around a grave or temple, damming a river to block the natural flow of water near a grave or temple, or mining a significant mountain.

III. QING EMPERORS ON MINING AND *FENGSHUI*

While the official law code excoriated some customs arising out of a concern for *fengshui* that were perceived to have pernicious social effects, such as delayed burials, the imperial state at the same time broadly recognized the classical *yin-yang* cosmology that underlined geomantic practice (Jones et al. 1994: 183). This section explores how Qing emperors and officials discussed *fengshui* in relation to mining. It shows that concerns about mining on the basis of *fengshui* did not begin on the local level and percolate up; in fact, the highest levels of the imperial state urged officials to be preemptively cautious about *fengshui* when opening mines and set out guidelines for dealing with concerns about mining.

By the late Qing, some elites looked upon the country's mining capacities with dismay. Mining seemed to be conducted neither "productively" nor extensively in many regions. Chinese intellectuals criticized the common people, and indirectly the state that appeased them, for deciding to leave valuable natural resources in the ground against what some came to see as in the economic interests of the country. An 1899 article in *Shenbao*, a leading newspaper published in Shanghai, captures an urban "reformist" sentiment regarding mining in the southwest:

The riches of China's mines are first amongst the five continents. . . . Foremost among [the provinces] is Sichuan, which has the better part of [the minerals] gold, silver, copper, iron, tin, coal, and lead, which are all stored in its mountains. In Sichuan, the mountains and hills are piled on top of one another and there are few flatlands; thus, it is suitable for holding many treasures. All of these minerals are stored here. It is a shame that my countrymen do not know that quarrying mines is for extracting profits; they let the natural profits from heaven and earth be abandoned in the ground, and they are apathetic and indifferent. They are deluded by the words of *fengshui* masters and say: "chiseling [the earth] breaks the veins of the earth and this will definitely hurt the *fengshui* [of an area]." Whenever they encounter a matter involving mining, they must try every way to prevent it, causing disasters ("Kai Shukuang yi," *Shenbao* 申報 1899.09.06, 01: 9481).

中國礦產之富，甲於五洲⋯⋯尤以四川為多‧大抵金銀銅鐵錫煤鉛諸礦，多蘊於山。蜀中峯巒層疊，絕少平原，宜乎韞寶甚多，而諸礦

無不畢備也。惜我華人不知開採，以濬利源，任天地自然之利，委棄
於地，而漠然無動於中 [sic][2]，甚且惑於堪輿家言，謂: "鑿殘地脈，必致
有碍風水"。 一遇開礦之事，必多方阻止，釀成禍端。

It is not my intention to ascertain whether *fengshui* hampered the Qing
economy; the overall effects of a particular cultural practice on economic
trends are challenging to measure. Thus, for the purposes of this article, I
will not address the accuracy of the aforementioned passage, which may
have invoked fengshui as a rhetorical strategy to advocate for a particular political position in the late nineteenth century. What I want to begin
with, rather, is the reason the writer of this article could even make such
an observation in the first place—that is, identifying *fengshui* as a reason
mining was not extensively pursued for "extracting profits." Although the
late Qing author blames "*fengshui* masters" for deluding the people, as
we will see further, protections for earth veins were mandated explicitly
through the edicts of Qing emperors from the beginning of the dynasty.
These edicts provided precedents for invoking *fengshui* to halt mining
activities.

The *Draft History of the Qing*, a semi-official history of the Qing
Dynasty, records the policy toward mining from the beginning of
the dynasty as, "If [mining] disturbs the *fengshui* of a forbidden
mountain, harms the field or graves of the people, attracts a mob
that disturbs the people, or if there is crop failure and [the price of
grain] increases, then ban [mining in the area]" [若有碍禁山風水，
民田廬墓，及聚眾擾民，或歲歉穀踊，輒用封禁] (Zhao 1977: 3664).
During the reign of the Kangxi Emperor (r. 1661–1722), the government established clear standards for the opening of mines, declaring:
"if there is an area with tombs, it is not permissible to mine [there]. If
a [mining activity] is occurring with issue, the provincial governor or
governor general should memorialize the throne to stop the activity"
[有墳墓處，不許採取。倘有不便，督撫題明停止] (*Zhongguo renmin
daxue Qingshi yanjiusuo* 1983: 5). The encouraging of bans on mining
activities around gravesites created a framework for officials to deal with
the disputes that inevitably arose from these extractive activities.[3]

In examining palace memorials sent to the capital from the seventeenth
century through the end of the dynasty, we notice a consistent pattern in
memorials about mining. When requesting permission to open a mine, governors and provincial officials explicitly stated that the *fengshui* of an area
was not affected by mining; when requesting a ban on mining in memorials, governors often explicitly stated that local gentry, commoners, or non-
Han peoples were concerned that the *fengshui* of an area was or would be

disturbed. In Qing discourse, mining was never thought to benefit a region's *fengshui*; mining could only hurt *fengshui* or have no effect on it. Even if a memorial concerning the opening of a new mine was not about the cosmology of a landscape, officials apparently felt compelled to address the issue as a matter of course at some point in at least one or two sentences, often toward the beginning of the memorial. This is not to say that all these disputes were primarily about *fengshui*; they had many different motivating factors. It is rather to say that *fengshui* was recognized as a discourse through which mining was often contested, and that officials could invoke *fengshui* to seek a mine's closures.

Let us look at some examples to appreciate just how widespread this practice was, even during a period of relative mining liberalization that characterized the Qianlong period. In the fifth year of the reign of the Qianlong Emperor (1740), Grand Academician and Minister of Rites Zhao Guolin (趙國麟) (1673–1751) memorialized the emperor to recommend opening coal mines in every province of the empire while suggesting the officials should allow the people to dig for coal freely as long as the mines do "not disturb the city moats, the dragon vein [of a town], or the tombs and mausoleums of the past emperors and the sagely worthies" [無關城池龍脈及古昔帝王聖賢陵墓] (*Zhongguo renmin daxue Qingshi yanjiusuo* 1983: 8). These guidelines, which were accepted and sent to provinces across the empire, required officials to first inspect the dragon vein and ancient graves of a locality before a mining site could be opened. In the following years, a string of governors composed memorials to the throne recommending bans be reinstated in certain regions of their provinces.

In 1743, the Governor General of Huguang petitioned for a ban as mining on the Miao borderlands. He related that some of the mines had been illicitly created in the "Miao Dominion" and disturbed the graves of local people (*Zhongguo renmin daxue Qingshi yanjiusuo* 1983: 16). Here, one could reasonably ask whether the Miao people practiced *fengshui*; while we might assume that some did while others did not, the imperial court assumed that the Miao people would share similar concerns and administratively anticipated such lawsuits. There is also the possibility that the imperial court was concerned about the *fengshui* of people who did not practice it themselves. In 1745, an official from Henan described how private mining had led "property-less" vagrants to migrate into the area, causing damage to the graves of the people (*Zhongguo renmin daxue Qingshi yanjiusuo* 1983: 14) In these instances, although the officials likely had many motivations to request an imperial ban, mentioning the disturbance of *fengshui* was routine practice.

There was a push and pull between the needs of economic development and claims made under *fengshui* that was remarked on by elites and
officials, who held a range of views on the subject. It was perhaps best
captured by the Qing statesman Zhang Tingyu (張廷玉) (1672–1755). In
advocating for increased mining operations, Zhang wrote: "If [a mine]
injures the dragon vein of an area, the people themselves will not mine
there; if [a dragon vein] is not disturbed, why should we leave the minerals in the ground" [若果有礙於龍脈者，人自不容其創取，其無關緊要
者，又何必棄之於無用?] (*Zhongguo renmin daxue Qingshi yanjiusuo*
1983: 12). Even for High Qing statesmen who advocated for more mining,
like Zhang Tingyu, this widely penetrating discourse had to be addressed
and negotiated; it was not window dressing. Reflecting this reality, the
gazetteer of Chenxi 辰溪 County in Hunan Province included the following observation in its mining section: "Our court is especially strong in
caring for the livelihoods of the people: [officials must say that a mine]
does not hinder [*fengshui*] and they must say it will not cause trouble"
我國家為民生計至尤極渥，然必曰並無妨礙，又必曰毋致生事   (Lin
2014: 231). Through the close of the late imperial period, both lucrative
mineral extraction and good *fengshui* could be taken as supportive of the
"livelihoods of the people." Though the two aims of mining minerals and
protecting *fengshui* could be in practical tension, neither were in ideological
contradiction with the goal of sustaining peoples' livelihoods.

As scholars have previously discussed, natural resource exploitation via
mining was substantial in the Qing, and my analysis here is selectively omit-
ting the many instances in which mining was allowed to proceed (Qiu 2001:
49–119). To be clear, *fengshui* was also invoked in many cases beyond the
topic of mining. A range of environmental phenomena could be cited as evidence for a decline of *fengshui*: a rise in tiger attacks in Huizhou's Wuyuan
County, prompted by the felling of forests and the mining of mountains,
was cited as indicative of a harmed dragon vein in the late Ming (Yi 2010:
135). The appearance of tigers—an inauspicious event—was rationalized
by the fact that locals had excessively felled forests and hence hurt the local
fengshui. But the main point here is to say that the discourse of *fengshui* penetrated to the very top of the Qing bureaucracy during discussions over the
livelihoods of the people. When officials asked the court to enact a ban on
mining, the court sometimes obliged; when officials petitioned to lift a ban,
the court also might oblige. If interpretation made *fengshui* appear like an
imprecise "superstition" to late nineteenth-century foreign observers, we may
point out the ironic flipside: it was precisely its interpretative nature that made
fengshui so powerful for the Chinese state to oversee a large economy while
negotiating the consequences of local development. To such ends, engaging *fengshui* in county courts worked for a preindustrial state that preferred

community negotiation, local preference, and regional discretion over blanket regulation from the metropole.

IV. LEGAL CASES FROM COUNTY ARCHIVES

Now that we have investigated the foundations of *fengshui* as an applied cosmology that saw land and health as link as well as the imperial state's engagement with geomantic discourse in official communications, this section turns to local lawsuits over mining. The following cases highlight instances wherein local gentry or commoners attempted to regulate or block intensive mining. Here, the information of the previous sections is woven together to show, first, how locals interpreted the effects of mining in their localities and, second, how the state dealt with their appeals.

Many disputes over mining occurred around Buddhist temples and Daoist abbeys. These religious sites were often located on prominent mountains, which could hold highly valuable minerals. Just like graves and houses, temples were ascribed with geomantic significance, as these were the edifices where the gods "lived" and around which dead were often buried.[4] In Ba County in Sichuan Province, official bans on mining were granted for Tiger Peak Immortal Temple in 1810 and again in 1819, when the monks of the temple attempted to exploit the mountain's coal deposits (*Sichuansheng dang'anguan* 1989: 280). When coal mining was said to have disturbed Dragon Chariot Temple and disturbed the *fengshui* of local graves, a magistrate responded: "for a long time officials have ordered this place forbidden (from mining), but Zhou Sicong et al. have the audacity to once again commence mining; this is truly obstinate, it is permissible to permanently expel him" [久經官斷嚴禁之處，周思聰等膽敢復興開挖，殊屬頑梗，准久逐] (*Sichuansheng dang'anju* 1991: 273). Similar appeals were made for White Cloud Temple (*Sichuansheng dang'anju* 1991: 274). Temples were particularly contested sites because their extensive mountainous properties were potentially lucrative for miners and because broader communities looked to temples as significant sites for protecting the local community.

In one petition to the Ba County court from 1838, a group of local gentry sought an official ban on mining around Crimson Silk Cloud Mountain (*Jinyun Shan* 縉雲山) and its namesake temple. In this case, the resident monks of the mountain had invited miners to dig for coal along the base of the mountain. The petition stressed the temple's prominence in attracting worshippers from both Ba and neighboring Bishan Counties, with the implication being that the desecration of the temple through mining would hurt the pilgrimage trade. The petition continued by ascribing the temple with the following cosmological significance:

But this temple is a mountain of immortals—one of the "nine great peaks and eight sceneries"—it is not only connected to the Literary Wind of both Ba County and Bishan County—but moreover it is part of the dragon vein of Yucheng (Chongqing). The preface of the inscription [of the temple] illuminates this evidence, and now it can be visited and examined. (Sichuansheng dang'anguan and Sichuan daxue lishixi 1989: 285)[5]

第該寺係原九峰八景仙山， 不但有關本邑、璧邑兩屬文風，抑且渝城系彼來脈，碑序朗憑，現可詣勘。

The gentry positioned their petition within two frames: first one drew on the religious geography of the temple and the second one explored its geomantic significance. If the mining continued in the vicinity of the temple, the gentry contended, the gods and sages would not be pacified. The second interpreted the temple as aligning along a flowing dragon vein that held great significance for the entire region around Chongqing.

Because the monks had colluded with the miners to allow quarrying around the site, the gentry were able to strategically invoke *fengshui* to express the wider community's concern for the environmental conditions of the temple against both the miners and the complicit monks. The gentry were likely empowered to do so through an awareness that a string of imperial edicts and regulations dating from the previous centuries had recognized the potential validity of such claims. Thus, the presiding Qing official was faced with the question of whether to permit the mining, which supported the monks' and the miners' livelihoods, or ban mining on the mountain. Because the magistrate came to identify mining along Crimson Silk Cloud Mountain as a trenchant, ongoing issue, having already created a string of lawsuits, he opted to ban mining on the mountain altogether (Chen 2019: 162).

The intended emphasis here is not to imply that cosmological interpretations of land were always accepted at face value by Qing officials. Qing officials neither said, to my knowledge, that dragon veins were not relevant, nor did they ever unilaterally ban all mining. A survey of mining-related archival lawsuits reveals that a great number of geomantic lawsuits exist, a great number of official bans exist, and a great number of rejected plaints exist—like any other type of petition, everything depended on immediate circumstance, the identities of the petitioners, presented evidence, and local precedence. Officials considered *fengshui* potentially relevant during litigation, but naturally preferred that local communities resolve issues outside of court. If endorsing an understanding of *fengshui* could prevent future litigation, officials seem to have been inclined to do it; if a commercial activity related to peoples' livelihoods had already occurred in an area for some time without issue, magistrates can be also seen expressing skepticism that sudden inauspicious events could be realistically traced back to mining activity.

The following case illustrates this last point precisely. In 1890, Deng Yulong (鄧愈隆), presented a plaint against his relative, Deng Xiaotong (鄧孝統).[6] Yulong noted that, in the past, members of the Deng lineage needed stones for construction of a road; Yulong resisted attempts to quarry stones on his land, but that Xiaotong and his gang persisted in their bullying. Then suddenly in the second month of the previous year, Xiaotong quarried a large section of the Azure Dragon Corner (*Qinglong zui* 青龍嘴) to build the road, which allegedly harmed the lineage's *fengshui* and caused inauspicious events to befall his household.

Xiaotong's branch of the lineage responded with a counter-plaint. They claimed that a total of three years had passed since the stones in question were quarried, and that each time stones were taken from the Azure Dragon Corner, which was considered commonly held lineage land, a banquet was held with all members of the lineage present. Xiaotong's side emphasized that the road would benefit the lineage considerably. They then proceeded to tell a very dark tale of the lineage's finances over the past months, wherein Yulong's branch had stolen the betrothal gifts of a fellow relation through a forged contract. Yulong's branch—one of the poorer families in the lineage—needed money to bury a recently deceased relative, which, in turn, spurred conflicts over the funds of the lineage's grave association.

There was no doubt that the Dengs had witnessed a difficult year. Marriages and deaths in the lineage had created new tensions over funding for burials, access to raw materials for building, and property shares in the communally owned Azure Dragon Corner. The magistrate did not address the financial questions of the lineage dispute as he likely had no way to determine whether the cited contracts were forged, but he did discuss the geomantic timeline. The first quarrying of the stones for the road construction had occurred three years previously; tensions built up over the subsequent years and reach a breaking point in the past year. This gave the magistrate an opportunity to disassociate the challenging events of the past year with the original quarrying three years earlier:

> The court rules ... even if [Yulong] claims that, in the past, stones were quarried from the Azure Dragon Corner behind his house, thereby obstructing the wind vein—three years have already passed [since the stones were quarried], and the two sides in this dispute constitute a single vein, they should not lose peaceful relations and engage in litigation. Now I order Deng Zhongchun (a witness and

relation) to privately meet with them individually to resolve the conflict, so that such incidents do not occur again. Everything is finished and the case is over; this is the ruling. [7]

堂諭...即稱以前打石挖伊宅後青龍嘴有礙風脈，已經三載，兩造
係屬一脈，不應失和興訟，著令鄧忠純私下與伊二比理息，
再無滋生事端，各結完案；此判。

In his ruling, the magistrate subtly pointed out two problems with Yulong's claims against his relatives. First, the magistrate noted that the two lineage branches in conflict constituted a single vein and their fortunes would be equally shared; resorting to formal litigation without first attempting mediation was not appropriate. Second, the Azure Dragon Corner had been originally quarried for stones three years previously, while unfortunate events (i.e., over marriage, health, and finances) had transpired only two years later. Here, the presiding magistrate drew on popular understandings of geomantic misfortune to rationalize the judgment: misfortune related to a grave infringement would likely occur within the lunar year of its occurrence—not three years later.

From emperors to commoners, historical archives suggest that nearly everyone was concerned about *fengshui* in some form in late imperial China. With increasing demands to open mines in light of the changing demographics of the empire, *fengshui* may have actually been elevated in both legal and political rhetoric alongside the "livelihoods of the people" to balance competing demands. *Fengshui* was certainly not consciously used to protect the environment, but it was invoked to sustain families, communities, and society at large. While it is likely impossible to reconstruct the precise material influences that the practice of *fengshui* had on the environment, we should keep in mind the aforementioned article from *Shenbao*, which along with many others published in the final years of the Qing observed that geomantic discourse was often invoked to keep resources in the ground. Local cases, just as imperial edicts before them, suggest this likely was the case in some instances.

V. CONCLUSION AND THEORETICAL IMPLICATIONS

It should be stressed again that there was no "environmental law" in form or in theory in imperial China, and *fengshui* was never intended primarily to protect the environment as a whole. As components of an ideology related to the conceptualization and construction of nature, geomantic veins in imperial times imbued landscapes with cosmological, legal, and political significance. Like other systems of cosmological spatial ordering, *fengshui* reflected existing ideologies and frameworks rather than any sort of self-conscious "environmental protection." In fact, it was through the widespread perception that *fengshui* helped secure wealth which enabled the discourse to be linked to political rhetoric over "livelihoods" in Qing times. This connection made applications of the cosmology potentially potent in both the imperial government and rural land disputes.

In particular, *fengshui* appears to have been tied up with notions related to what we could broadly term "sustainability" by enabling the preservation of certain resources and landscapes over time for the sake of social stability. Officials in practice recognized *fengshui* as a negotiating mechanism in local courts to permit occasional checks on development in the wake of the commercialization of rural land markets that accompanied the growth of the Qing population.

As the number of works on environmental history has increased, scholars have begun to move away from linear narratives of environmental decline in China, revealing in the process that while environmental change was constant, adaptation, sustainability, and even recovery were at times part of the story as well. For the Qing period in particular, scholarship has now investigated the ways in which notions of conservation and landscape management were actively voiced and debated across the bureaucracy and broader society (Bello 2016; Duara 2017: 65–86; Schlesinger 2017). A desire by Manchu elites to protect the dragon vein of their ancestral homeland may have contributed to some of the policies noted by Jon Schlesinger and David Bello. This chapter has suggested that similar frameworks were found on smaller scales across much of the empire, with the imperial bureaucracy seeking to sustain elements of natural landscapes that made intensive agriculture possible while preserving the social harmony of communities responsible for farming them. In other words, the late imperial state appears to have been politically invested in the durability of certain agricultural and natural landscapes.

And yet, even as the study of environmental history in China has never been more popular, serious studies of *fengshui*—the dominant discourse for actually discussing and addressing landscape matters throughout the history of late imperial China—remain exceedingly few. This lacuna is understandable, in that *fengshui* did not protect the environment as much as it reflected and sustained the social order through providing a language, mobilized constantly in law and politics, through which people could attribute fortune and misfortune to a dynamic relationship with the natural and constructed world around them. *Fengshui* seemed to lack a readily identifiable label within much of twentieth-century scholarship. Yet there has been a growing interest in "Traditional Ecological Knowledge," defined as "a cumulative body of knowledge, belief, and practice . . . handed down through generations through traditional songs, stories and beliefs," which concerns "the relationship of living beings with their traditional groups and with their environment." This concept nicely accords with at least some elements of what *fengshui* was in imperial times.

China's long imperial history was marked by ecological decline—a process that Mark Elvin famously termed "the retreat of the elephants" (Elvin 2008). What this chapter has suggested is that, in spite of that decline, there were mechanisms and languages available to people in China to understand changes in their local environments and mobilize to prevent changes that were deemed unwanted. During the Qing, people walked for miles to reach courts to present lawsuits on the harming of *fengshui* because they saw elements of the natural world as dangerous, threatening, and ultimately, highly relevant to their lives. In the resultant trials and memorials, we may appreciate the pervading sense of the great natural fragility of the cosmic order that was perceived by emperors, officials, and commoners alike. Qing legal sources invoking *fengshui* provide a large documentary record of human thought that envisioned all possible events as potentially driven by the natural and built environment, and one that was once recognized through law. In corners across China up until the end of the imperial period, some forests—many anthropogenic, but also a few natural ones—survived, some mountains remained un-mined, and some rivers ran undammed at least in part because there were socially acceptable ways to articulate cogent legal and political arguments for maintaining them as such.

NOTES

1. For a history of this term, see Margherita Zanasi, *Economic Thought in Modern China: Market and Consumption, c. 1500–1937* (Cambridge: Cambridge University Press, 2020), 16–50.

2. This phrase *wudong yu zhong* 無動於衷 is today often written with the character *zhong* 衷 rather than the character *zhong* 中.

3. It is generally held that the Yongzheng Emperor (r. 1722–1735) limited mining in the southwest, while the Qianlong Emperor loosened this policy. Mining bans were commonly issued through the end of the dynasty, and certain provinces, such as Yunnan, were sometimes given exceptions to bans due to the importance of its copper and silver mines for the empire's currency. For a discussion of Yunnan's mining industry, see Yang Shouchuan 楊壽川, *Yunnan kuangye kaifashi* 雲南礦業開發史 (Beijing: Shehui kexue wenxian chubanshe, 2012).

4. Monks were frequently accused by local gentry of hindering the *fengshui* of temples. See for instance: Sichuansheng dang'anju 四川省檔案局, ed. *Qingdai Sichuan Baxian yamen Xianfengchao dang'an xuanbian* 清代四川巴縣衙門咸豐朝檔案選編, vol. 9 (Shanghai: Shanghai guji chubanshe, 2011), 9.

5. The full record of this case in the Ba County Archive is scattered over four case files. For these files, see Ba County Qing Archive: 6.15.17045, 6.15.17095, 6.15.17103, and 6.15.17133. I thank Gilbert Zhe Chen for directing me to these sources.

6. Nanbu County Qing Archive: 10.662.01.
7. Nanbu County Qing Archive: 10.662.07.

CHINESE LANGUAGE SOURCES

Anonymous. 1899. "Kai Shu kuang yi" 開蜀礦議 ["An Opinion on Opening Mines in Sichuan"]. *Shenbao* 申報, vol. 9481.

Langzhong Municipal Archive 閬中市檔案局. *Dili yuanxu* 地理原序 [*Preface to the Origins of Earthly Principles*]. Baoning, Sichuan. c. 1875-1912.

Lin, Rongqin 林荣琴. 2014. *Qingdai Hunan de kuangye: fenbu, bianqian, difang shehui* 清代湖南的礦業：分布·變遷·地方社會 [*Mining in Qing Hunan: Distribution, Transformations, and Local Society*]. Beijing: Shangwu yinshuguan.

Qiu, Pengsheng 邱澎生. 2001. "Shiba shiji Diantong shichang Zhong de guanshang guanxi yu liyi guannian" 十八世紀滇銅市場中的官商關係與利益觀念 ["'Interests' in Economic Organization: The Shaping of The Yunnan Copper Market in Eighteenth-Century China"]. *Zhongyang yanjiuyuan lishi yuyan yanjiusuo jikan* 中央研究院歷史語言研究所集刊, vol. 72, no. 1, pp. 49–119.

Sichuansheng dang'anguan 四川省檔案館 and Sichuan daxue lishixi 四川大學歷史系, eds. 1989; 1996. *Qingdai Qian Jia Dao Baxian dang'an xuanbian, shangxia* 清代乾嘉道巴縣檔案選編 (上下) [*Selections from the Qing Dynasty Archives of Ba County from the Qianlong, Jiaqing, and Daoguang Reigns, Volumes One and Two*]. Chengdu: Sichuan daxue chubanshe.

Sichuansheng dang'anju 四川省檔案局, ed. 1991. *Qingdai Baxian dang'an huibian, Qianlong juan* 清代巴縣檔案彙編，乾隆卷 [*A Compendium of Qing Archives from Ba County, Qianlong Collection*]. Beijing: Dang'an chubanshe.

Sichuansheng dang'anju 四川省檔案局, ed. 2011. *Qingdai Sichuan Baxian yamen Xianfengchao dang'an xuanbian* 清代四川巴縣衙門咸豐朝檔案選編 [*Selection of Qing Archives of the Xianfeng Reign from Sichuan's Ba County Yamen*], Volume Nine. Shanghai: Shanghai guji chubanshe.

Wu, Baosan 巫寶三 and Li Puguo 李普國, eds. 1990. *Zhongguo jingji sixiangshi ziliao xuanji: Ming Qing bufen* 中國經濟思想史資料選輯: 明清部分 [*Selected Works of the History of Chinese Economic Thought: The Ming and Qing Dynasties*]. Beijing: Zhongguo shehui kexue chubanshe.

Yang, Shouchuan 楊壽川. 2012. *Yunnan kuangye kaifashi* 雲南礦業開發史 [*A History of the Development of Mining in Yunnan*]. Beijing: Shehui kexue wenxian chubanshe.

Zhao, Erxun 趙爾巽. 1977. *Qingshigao* 清史稿 [*Draft History of the Qing*], Vol. 124. 1928 Rpt; Taibei: Zhonghua shuju.

Zhongguo renmin daxue Qingshi yanjiusuo 中國人民大學清史研究所, ed. 1983. *Qingdai de kuangye, shangce* 清代的礦業, 上冊 [*Mining in the Qing Dynasty, Vol. One*]. Beijing: Zhonghua shuju.

Ziming, Chen 陳自明. 2007. *Furen Daquan liangfang* 婦人大全良方 [*Complete Good Prescriptions for Women*]. Beijing: Zhongguo zhongyiyao chubanshe.

ENGLISH LANGUAGE SOURCES

Bello, David A. 2015. *Across Forest, Steppe, and Mountain: Environment, Identity, and Empire in Qing China's Borderlands.* Cambridge: Cambridge University Press.

Bennett, Jane. 2009. *Vibrant Matter: A Political Ecology of Things.* Durham: Duke University Press.

Berkes, Fikret. 1993. "Traditional Ecological Knowledge in Perspective." In *Traditional Ecological Knowledge: Concepts and Cases*, edited by Julian T. Inglis, pp. 1–11. Ottawa: International Development Research Center.

Brook, Timothy. 2010. *The Troubled Empire: China in the Yuan and Ming Dynasties.* Cambridge, MA: The Belknap Press of Harvard University.

Bruun, Ole. 2003. *Fengshui in China: Geomantic Divination Between State Orthodoxy and Popular Religion.* Honolulu: University of Hawai'i Press.

Chakrabarty, Dipesh. 2009. "The Climate of History: Four Theses." *Critical Inquiry*, vol. 35, no. 2, pp. 197–222.

Chen, Gilbert Zhe. 2019. "Living in This World: A Social History of Buddhist Monks and Nuns in Nineteenth-Century Western China." Washington University in St. Louis PhD Dissertation.

Duara, Prasenjit. 2017. "The Chinese World Order and Planetary Sustainability." In *Chinese Visions of World Order: Tianxia, Culture, and World Politics*, edited by Ban Wang, pp. 65–86. Durham: Duke University Press.

Elvin, Mark. 2008. *The Retreat of The Elephants: An Environmental History of China.* New Haven: Yale University Press.

Jones, William C., Tianquan Cheng, and Yonglin Jiang. 1994. *The Great Qing Code.* New York: Oxford University Press.

Miller, Ian M. 2020. *Fir and Empire: The Transformation of Forests in Early Modern China.* Seattle: University of Washington Press.

Perdue, Peter. 2010. "Is There A Chinese View of Technology and Nature?" In *The Illusory Boundary: Environment and Technology in History*, edited by Martin Reuss and Stephen H. Cutcliffe, pp. 101–119. Charlottesville: University of Virginia Press.

Rowe, William T. 2001. *Saving the World: Chen Hongmou and Elite Consciousness in Eighteenth-Century China.* Stanford: Stanford University Press.

Schlesinger, Jonathan. 2017. *A World Trimmed with Fur: Wild Things, Pristine Places, and the Natural Fringes of Qing Rule.* Stanford: Stanford University Press.

Yi, Qiufang. 2010. "The Dragon's Veins: Public *Fengshui* in Late Imperial Wuyuan County." Syracuse University Doctoral Dissertation.

Zanasi, Margherita. 2020. *Economic Thought in Modern China: Market and Consumption, c. 1500–1937.* Cambridge: Cambridge University Press.

Zhang, Meng. 2021. *Timber and Forestry in Qing China: Sustaining the Market.* Seattle: University of Washington Press.

Chapter 4

Grave Matters

Geomancy and Neo-Confucian Resistance to Grave-Removal in Central China

Liang Yongjia[1]

In 2012, the Henan provincial authorities launched the infamous Henan Grave-Removal Campaign (河南平墳運動), aimed at removing millions of graves scattered throughout the province's rural areas. The campaign was enforced in four prefectures, and implemented with particular strictness in Zhoukou (周口). The policy demanded not only that all newly deceased residents undergo cremation but that all existing graves should be "leveled" (*ping* 平) through tomb-top and tombstone removal. The campaign caused a nationwide outcry that ultimately led to direct intervention from Beijing, leaving the Henan authorities no choice but to abort the effort.

Scholar-activists deemed the Henan Grave-Removal Campaign one of the top-ten "civil society incidents" of the year.[2] National and local elites denounced the campaign through hundreds of online protests. Such protests were widespread at that time, but unimaginable now in 2021. Yet the story's persistence deserves scrutiny primarily because it suggests a continuation of Chinese traditions. Paradoxically, the narrative of these protests persists and has even been augmented in the present day as a narrative representative of elitist Neo-Confucian understandings of the environment. Of the polyphonic array of condemnations surrounding the project, one of the more interesting critiques came from a group of self-identified Neo-Confucianists whose spontaneous expression of Chinese religious beliefs dovetailed with local people's concerns about geomancy—a silent but resilient reason for villagers' reluctance to remove the graves. Whether consciously or not, in their compromising of the government's developmentalist policies, both elite Confucianists and grassroots geomancy practitioners can be understood to hold values about the dead. These different concerns about the care of the dead together resisted the environmental damage that would have resulted from such unpopular campaigns.

Prasenjit Duara (2015: 4) suggests that "Asian traditions," and especially Confucianism, present "hope" for mitigating environmental disasters brought about by the "radical transcendence" of Abrahamic religions and their renewal in Protestantism in the modern West. According to Duara, Protestantism gave rise to the conflation of faith-based community with the emerging nation-state, based on cultural and religious identities. The emphasis on radical singular religious Truth meant that, as these modern derivative nation-states became secularized, they subscribed to the new transcendent Truth-drives of "progress" and developmentalism, aided by a linear history. Thus, Protestantism gave rise to modern nationalisms, which were not only intolerant of alternative cultures, but in their support and guidance of capitalist developmentalismwere in many ways responsible for the global environmental crisis (Duara 2015: 160–162).

As a homogenizing process clustering one denomination by removing others, confessionalization—the process in which identical denominations congregate territorially, excluding other churches—became the standard for nation-state formation in Asia, when the nationalistic elites departed from their civilizational frameworks in order to build states based on "tunneled histories," by which the nation was authenticated as a linear development from ancient times to the present, without significant influence from neighbors. Duara (2015) has noted the ability of non-Abrahamic religions such as Confucianism, Buddhism, Daoism, and Hinduism to recognize each other's transcendental values, and he calls them examples of "dialogical transcendence". It is antithetical and complementary to the "confessional communities" or "radical transcendence" of Abrahamic religions and the modern nation-state. Therefore, we see the coexistence of both radical and dialogical transcendence in Asian countries' nation-building. This is even more true in China of 2000s and early 2010s, where thousands of NGOs for environmental protections operate effectively by balancing ecological concerns with the developmentalist national and local governments (Duara 2015: 239–278). In this light, Neo-Confucianist condemnation against the grave-removal campaign was probably one of the hopes Duara put forward to remedy the "crisis of global modernity."

Duara's contrast between Asian dialogical and Western Christian radical transcendence makes sense if one moves beyond the academic habitus of deconstructing every super category by dismissing it as essentialism. After all, modernity constitutes a unique and novel way of thinking not found in the non-Western world. A century after the Reformation, "naturalism" began in the West. It soon spread to Asia, reaching China in the late nineteenth century when elite Chinese reformers started to embrace many Western discourses of science, evolutionism, and the idea of an objective "nature." Philippe Descola (2013) traces Western modernity back to the intellectual objectification of "nature," as reflected in the paintings of Roelant Savery. According to Descola, the watershed moment where "nature" became distinctively different from "culture" occurred sometime during the seventeenth century. He refers to this time as a transition from full-fledged era of "analogism" to "naturalism"; "analogism" is "a mode of identification that divides up the whole collection of existing beings into a multiplicity of essences, forms, and substances separated by small distinctions and sometimes arranged on a graduated scale so that it becomes possible to recompose the system of initial contrasts into a dense network of analogies that link together the intrinsic properties of the entities that are distinguished in it" (Descola 2013: 201). Descola defines objectified nature as "naturalism," or "the continuity of the physicality of the entities of the world and the discontinuity of their respective interiorities" (Descola 2013: 173). The transition has also been established by other contemporary thinkers in a similar vein. Charles Taylor argues that Deism in the seventeenth century took a naturalistic turn when it developed toward exclusive humanism by reifying the idea of an objective nature distinct from humans (Taylor 2007: 222–224). What followed after naturalism is a well-known story of how nature became external to humans; demoralized, manipulatable, and ultimately vulnerable to human activities. In a way, the shift toward naturalism can be conceived of as marking the start of an irreversible "Anthropocene."

In this chapter, I will discuss these issues as they relate to the Henan Grave-Removal Campaign. Imported "naturalism," which was very ironically translated into Chinese using Laozi's concept of *ziran* (自然), has penetrated the "economic developmentalism" of the modern Chinese state. I argue that the campaign's real purpose was to transfer rare and valuable reclaimed arable land to major metropolises within the province for purposes of industrial development. I then probe the real reasons behind the Zhoukou villagers' reluctance to remove the graves of their deceased parents and ancestors. Resistance involves deep-rooted beliefs in geomantic ideas about the proper flows of "wind" and "water" (fengshui, 風水) that a stable, harmonious environment requires. In the third section, I analyze the opinions of the self-identified Neo-Confucians who openly condemned the campaign by elaborating on the dichotomy that exists between the government's developmentalist policies and values of ancestor worship commonly held by "the Chinese people." By dichotomizing the Chinese state with "the people" in terms of religion, the Neo-Confucians attempted to rescue Chinese tradition from a utopian, radically transcendent form of global developmentalism emerging from the more "Westernized" contemporary Chinese state.

I. ENVIRONMENTAL DEGRADATION AND THE CAMPAIGN TO REMOVE GRAVES IN HENAN

The Henan Grave-Removal Campaign began under the auspices of Lu Zhangong, Secretary of the Henan Provincial Committee of the Communist Party of China. During his inspection of Zhoukou Prefecture in November 2011, the top leader of the province commented on the many tomb-tops and gravestones visible throughout the region's crop fields. At the end of the month, the prefectural CCP secretary, Xu Guang, acted on Lu's remark by moving to "reform" burial practices. He placed reform as the top agenda for the prefectural Bureau of Civil Affairs and began formally implementing it in a prefectural decision-making meeting on February 22, 2012. A week later, Mayor Yue Wenhai (岳文海) of the prefectural seat gave an address on the importance of removing the tombs. He stated that the "extravagant funeral expense" had become a burden for the "social masses," who longed to rid themselves of "ignorant superstitions." Above all, he emphasized the loss of arable land to burial sites, stating, "According to incomplete statistics, there are over three million graves [in the prefecture], occupying about 35 thousand *mu* or 23.3 square kilometers.[3] At a rate of 6‰, about 60 thousand people will die each year, taking away as much as over 600 *mu* of arable land [annually] . . . . The graves have caused a significant loss of arable land, seriously jeopardizing social demands and hindering rural economic and social development. Reforming the burial practices would benefit our people." He concluded by stating that the need for reform "has reached a critical point, leaving us no choice."[4]

Similar decrees from high leaders were consolidated into a formal document, *Remarks on Further Steps to Reform Burial Practice*. The document was issued as a joint release from the Chinese Communist Party Committee and the People's Government of Zhoukou Prefecture in March 2012. It specified three goals: mandatory cremation for the deceased in rural areas, and full coverage of "public welfare columbaria"; a complete ban on land-burial using coffins or urns; a ban on the installation of new tomb-tops; and the gradual removal of existing tomb-tops. Shangshui County was singled out as the first site to implement the campaign as part of a three-year program.[5] The project was initiated with an assembly of thirty thousand people and mass propaganda. Foreseeing the unpopularity of the campaign, the local government hired military veterans and formed them into a "Law-implementing Team for Burial Reform," equipped with sixty-five vehicles and forklifts.[6]

Several conspiracy theories suggested that the campaign had dark intentions. One was that it reflected the personal anger of the contemporary top provincial leader, who felt his political career was in decline. Another theory suggested that the campaign was meant to destroy the geomancy of

China's next leader, whose ancestors were buried in the province. Removing graves would alter the landscape and, ultimately, damage the "dragon's vein" (*longmai*). Though the campaign was indeed radical and there were many silent adherents to geomancy (as will be discussed in the next section), these conspiracy theories are not fully believable. In the Chinese political system, it would be career suicide for a provincial head to unleash personal anger in public policy or attempt to attack his future boss with sorcery. The project's instigator, Lu, was safely ensconced in an honorable position—the vice-speaker of China's People's Political Consultative Conference —and thus able to dodge the repercussions of the project. However, other leaders did not fare as well. Apart from the demotions of several officials involved in the campaign, Mayor Yue eventually stepped down. A now-disgraced politician, he later assumed a position in a shareholding company. Following an August 2019 investigation, Secretary Xu Guang, the then top leader of Zhoukou Prefecture who implemented the campaign, was expelled from the Party and arrested on corruption charges. In other words, they were not punished for their involvement in the campaign. Given the unenviable fate of these officials, the conspiracy theories are not persuasive.

At least one believable reason for the unpopular campaign was rather apparent. The provincial authority was desperately in need of land, not only for agriculture but for industrial development.[7] Until 2010, the urbanization rate in Henan was merely 37.7 percent, ranking at the bottom of the country a whole 12 points lower than the national average. By these figures, failure to attract industrial investment had left the province severely "underdeveloped" and lagging behind other provinces in terms of its economic capacity. In an era when government expenditure depended mostly on local revenue, and political careers depended on the success of an official's economic policy, the Henan leadership was under enormous pressure to convert any land it could from agricultural to industrial use. However, the province was a major agricultural base, and in order to safeguard the country's "red line" of 180 million *mu* of arable land, the State Council (國務院) led by pro-agriculture Premier Wen Jiabao had rejected several petitions from the provincial authority for creating industrial parks. The most significant initiative rejected was the "California Industrial City," a 2006 development project agreed upon between Shi Jichun, then Governor of Henan, and Arnold Schwarzenegger, then Governor of the State of California. Thus, the only remaining hope seemed to be creating new arable lands so that they could transform old functions of land into industrial use. Yet Henan is a flat landmass where agricultural activities have been carried out uninterrupted for over four thousand years. As the province simply had no land to reclaim, removing graves emerged as a promising method for land reclamation.

Industrial development was a carefully planned move on the part of the government. In fact, as early as 2009, the provincial authority had already come up with a plan which effectively removed obstacles to translating grave sites into industrial land. In a document entitled *Temporary Trial Measures for the Principle of "On-off Correspondence" in the Use of Urban-rural Land for Construction Purposes*, the provincial government made allowances for the permutation of industrial land on the condition that an equal amount of rural land were reclaimed for agricultural use. In September 2011, the State Council approved Henan's request for exploring the possibility of "human-land correspondence," which would allow for the inter-prefectural transfer of land quotas for agricultural use. In other words, if parcels of rural land allotted for construction were reclaimed for agricultural use in a non-industrialized prefecture such as Zhoukou, the area's "quota" could be "sold" to an industrialized city like Zhengzhou (鄭州), the provincial capital. Such transfers would dramatically increase the marginal profits of reclaiming rural land, thereby strongly incentivizing the Zhoukou government to remove the graves. "We are eagerly looking forward to the details of the 'human-land correspondence' policy, and can't wait to implement it," said one official in the Zhoukou prefectural government at the time.[8] It is little wonder that only one and a half months after the State Council approved the policy, the provincial head initiated the campaign by commenting on the tomb-tops in Zhoukou.

In fact, the practice of removing graves was not new. In 1958, a national campaign to "grab crops from ghosts" was enacted during the Great Leap Forward. At the time, the state accused Chinese burial practices of being "outdated" and campaigned for the removal of old graves (Bruun 2003: 87). Official propaganda suggested the Chinese should bury the dead deep in order to save arable land. That year, Huang Yanpei, the vice-speaker of the National Congress, set an example by voluntarily digging up his ancestors' graves. His action was followed by a declaration by CCP leadership, who unanimously vowed to be cremated without graves after their deaths. Grave-removal became even more radical during the Cultural Revolution when many tombs and cultural relics were destroyed for their association with "the Four Olds." Cremation challenged traditional Chinese values which called for proper burial of the deceased, as well as regular visits to, and maintenance of graves as essential obligations. Furthermore, removing graves ran the risk of ghosts haunting the village (Whyte 1988).

Complicating the matter was the fact that, unlike earlier grave-removal campaigns, the Henan campaign was not intended to recoup land for agriculture, but for industrial use. Removing tomb-tops to spare land for industrial use thus suggested the arrival of more factories, from high-pollution industries like paper mills, leather factories, chemical plants, and coal processing plants. The campaign was launched at a time when industrial environmental degradation was already front and center in the public spotlight. Severe haze was consistently overwhelming major metropolises and industrial centers such as Beijing, Shanghai, Tianjian, Chongqing, and Shijiazhuang, but cities in Henan were arguably among the hardest hit. Post-2004, a series of water pollution events in Henan alerted the public to the hazards of industrial pollutants. Across the province, surface water and air quality ranked among the lowest on the national environmental index. In some cities in Henan, water and air levels were even found to be below the minimum requirement for human consumption.[9] Zhoukou Prefecture was among the worst polluted areas in the whole country in terms of air, water, and soil. Despite having very little industry, the prefecture shares its air and river with neighboring industrialized centers. Zhoukou is also host to many labor-intensive mills and factories unable to survive the stricter anti-pollutions regulations in place in China's metropolises. It was reported that, since 1997, many ponds in Zhoukou had become so "black and stinky" that fish in the ponds often died of pollution, or "turned upside down" (*fantang*), in local parlance.[10]

In September 2007, an outspoken national newspaper, *China Youth Daily*, reported that in a Zhoukou "cancer village" of over 3,000 villagers, 704 people had died of various cancers in about fifteen years. It was believed that severe pollution of the Ying river was the primary cause. Heavily polluted by dangerous substances such as Nitrate nitrogen and manganese, the river running through the prefecture was technically no longer a river of water, but of chemicals. The pollutants came not only from the local MSG factory, but from hundreds of small leather factories and plastic plants that were deliberately discharging heavy metals and hazardous chemicals directly into the river at no cost. In 2013, China Central Television even reported that farmers in a Henan village had been using wastewater from a paper mill to grow wheat. One farmer stated that the wheat had to be sold outside the area because no local people would dare to eat it.[11] Despite their constant health issues and life-threatening situations, villagers found the local officials more interested in their careers than improving the environment. One villager complained about "you"—the officials, stating, "You got benefits, we got sick; you earned a promotion, we earned a passage to Heaven (*i.e.*, death)" (你們得利, 俺們得病; 你們升遷, 俺們升天).[12]

While there was little doubt that the grave-removal campaign would undoubtedly worsen industrial pollution, in November 2012, the Zhoukou prefectural government declared they had restored 30,000 *mu* (about 20 km²) of agricultural land by removing two million tomb-tops. The quota had been handed to the prefectural Bureau of Land Resource, and was awaiting transfer to the provincial authority.[13] In a speech celebrating the achievement of the campaign, provincial boss Lu Zhangong restated the merits of the effort by commenting that the campaign adhered to the famous Hu Jintao's overarching policy of "a scientific outlook on development" (*kexue fazhanguan*).[14] Doubts circulated as to whether the numbers had been exaggerated, as two million tomb-tops, many of which should have already been categorized as agricultural land, could not possibly amount to 20 km². Removing them was more a restoration of the agricultural land than new reclamation. There was also the fact that a good number of topless burial sites remained untouched as villagers endeavored to refrain from growing anything on the graves of their deceased parents.[15] The newly presented land quota figure for industrial use was thus largely an abstract number generated through statistical and bureaucratic magic. Either way, its environmental implications were clear. Though grave-removal had not as yet yielded much spare land, now, polluting industries were to have a new back door.

The Henan campaign was as developmentalist as it was utopian. It was developmentalist, in the sense that the campaign reflected the desperate situation the provincial government faced in creating room for industrial development at the expense of severely damaging the local people's social fabric. Yet the campaign was utopian too. By destroying the physical presence of the graves, it inherited and continued the communist imagination of instant deliverance into a blissful materially prosperous future, one that was disconnected from the present by cutting its relations with the past—the deceased relatives and ancestors. As a recurring social theme, grave-removal was thus meant to remove the individual and kinship identities of the villagers and atomize them in support of national development. At the same time, the dual nature of the grave-removal campaign intensified the dichotomy of culture and nature. The fact that desecrating graves was even considered as an option reveals that the province was already in an environmental crisis—there was no more land to reclaim, no more lakes to drain, no more trees to fell, no more wetlands to fill. Indeed, Henan provides a classic example of agricultural involution, by which, no matter how labor-intensive the economy, the only way to ensure its sustenance is to abandon it in favor of industrial development and urbanization. With this campaign, the local villagers were not just objectified for government economic development purposes, but they themselves had become culturally barren in the sense that they had to face a landscape without ancestors.

Mayfair Yang (2008) solidly establishes how the genealogical study of Chinese history under Maoism came to reflect a Chinese version of "unilinear social evolutionary history." This social evolutionary approach to history was proposed by Marx and Engels' understanding of Lewis Henry Morgan, and by Joseph Stalin in his exposition of the "dialectical and materialist history." Yang's discussion of social evolutionism's hegemonic role in promoting the social engineering projects of the Communist Revolution helps us understand how such a campaign could be at once materially developmentalist and spiritually utopian. In reality, these two angles are nothing more than two sides of the same coin—a coherent conception of linear evolutionary history.

II. ENVIRONMENTALLY FRIENDLY: THE SILENT RESISTANCE OF GEOMANCY

Though the goal of the Henan Grave-Removal Campaign was to squeeze natural resources, there was social resistance to the prospect of making the dead die again for the sake of the living. Aware of the unpopularity of the policy and possible backlash, some local authorities in Henan promised a 200 CNY (about 30 USD) reward to those who voluntarily leveled their family members' graves.[16] Various local government bodies promised higher rewards (600 CNY) or charged 580 CNY for installing urns in public columbaria. Local officers who failed to fill required quotas would be demoted or expelled from office. Village heads, government-paid teachers, and party members were all instructed to remove their family graves, or be deprived of their positions. In some counties, peasants who refused to remove graves were to be barred from accessing social securities. Veterans where even hired to back the policy by force as prefectural and provincial leaders pressed their subordinates to accept that "there was no turning back" and "the only way is to proceed till the mission was accomplished."

It was no surprise that the campaign triggered furious protests. In an already chaotic 2012, politicians, official propaganda machines, intellectuals, the liberal press, and netizens were up in arms. In May, a legislator in the provincial congress, Zhao Keluo (趙克羅), revealed unequal measures of the campaign in his microblog by demonstrating that graves connected to officials ranked higher than "deputy prefectural chief" (*fuchuji* 副處級) enjoyed exemption from removal.[17] He immediately came under fire from the provincial authority and even published an apology for spreading "false information." He later released a death note stating that he was "ready to give up his life for justice," and darkly implying that he might have to commit suicide.[18] It is worth noting, that his first microblog was problematized as the government had publicized a plan to "protect, relocate, or maintain the special graves of famous people." However, as has been pointed out, the category of "famous people" (*zhiming renshi* 知名人士) is a malleable term which "any official can easily claim for his ancestors."[19]

If it were not for a peasant couple who were crushed to death by a tombstone that they had voluntarily toppled in October 2012, the campaign likely would have been carried out till "the goal was met."[20] Following this event, major official news outlets like *Guangming Daily* (光明日報), *Phoenix TV*

Network (鳳凰衛視), and *China Youth Daily* （中國青年報） began to provide full coverage of the campaign and offer severe criticisms about its inconsistency and unfairness. A turning point came on November 16, 2012, when the State Council released Decree 628, revising Article 20 of *Regulations on Mortuary Practices* by deleting a provision, which allowed local Departments of Civil Affairs to remove graves by force. The decree was crystal-clear: by depriving the local government of the coercive power to destroy graves, the state wanted the campaign to stop.[21] A spokesman from the Ministry of Agriculture declared in December that the campaign was one of "good intentions, but the specific measures were not appropriate."[22] Zhoukou and other prefectural authorities in Henan seemingly had no choice but to stop their campaign. The halt was silent, with people only finding out that the campaign no longer existed much later, in February 2013.

One may wonder how removing graves in a single province become a matter of national controversy. Beyond politically directed rallies against inequality and corruption, there were also the subjective realities of ordinary villagers attached to the project. One of my most perplexing discoveries is the fact made clear from interviews and news reports that the local peasantry was not strictly opposed to leveling tomb-tops and tombstones. In fact, the issues they were most opposed to were cremation and the relocation of urns to public columbaria.

Reluctance to accept these facets of the campaign involves a silent, deeply embedded philosophy informing Chinese burial practice: *fengshui* (風水). Literally, "wind and water," *fengshui* is often translated as "geomancy." Geomancy is a complex body of specialized knowledge guiding the building of houses and graves, and other settlements too. It is one of the most popular and ancient Chinese beliefs still observed today and possibly the broadest religious practice among Chinese-speaking people worldwide (see Feuchtwang 1974; Bruun 2003). As defined by Mayfair Yang (2004: 731), *fengshui* "is an ancient art or technology to improve people's physical and spiritual lives by aligning the buildings in which they live and the graves in which their ancestors are buried, to harmonize with the flow of 'primordial energy.' "

As a non-institutionalized set of beliefs and practices, *fengshui* is also perhaps the most tolerated form of popular religion in post-1949 China (Bruun 2003: 82). It experienced a particularly strong revival after the Cultural Revolution. Ethnographic studies of post-Mao China have shown that *fengshui* is not just widely practiced, but a highly important cultural informant when it comes to burials. Urban or rural, central or peripheral, rich or poor, communities throughout China generally adhere to *fengshui* in their burial practices. Henan is no exception, and a *fengshui* master is a necessary figure in any ordinary Henan funeral. A *fengshui* master, referred to in Henan as a "specialist of houses and graves," specializes in identifying proper sites for houses (*yangzhai*) and graves (*yinzhai*). He provides in-house ritual services to "customers" who almost always seek expert opinions about the burial of their deceased parents (Chau 2019).

A *fengshui* master's most important skill is ensuring the burial site receives proper flows of "wind" and "water" from its surroundings. A master from Henan informed me that an ideal burial site is one where the grave "faces the river and lean against a mountain, where wind and water constantly flow around for the dead to catch." The dead should be able to see as great a variety of landscapes as possible, while resting in a hidden and quiet place. One has chosen a good burial site if, after the burial, grasses grow fast on and around the graves—a sign of prosperity. In pre-1949 China, there were competitions among the rich for good burial sites, on terraces, in woods, along riverbanks, and by roadsides, because these places were believed to host constant flows of wind and water. Most lineages like to bury their ancestors at the same site in what are called "ancestor graves" (*zufen* 祖墳) in order for the living to enjoy the same kind of prosperity blessing. Paying homage to the ancestral graveyard is a common practice for descendants at the time of important events like the national examination or in times of illness among the elderly. "I often went to the graveyard for my ancestors' blessing when my son was in middle school. He is now in college," a father from the Zhou lineage said.[23]

Conviction in the efficacy of ancestral blessing to the descendants' exam success because of auspicious *fengshui* is very deep. According to a junior professor I interviewed, his family's ancestral graves in a Zhoukou village became the object of envy of his fellow villagers after he was admitted to Peking University. In the first few months after he left for Beijing, his parents slept by the graveyard in fear that jealous neighbors would try to discreetly destroy the family graves. He was puzzled by the local government's campaign, stating, "It would be very clear to them what it means to remove graves. I guess they don't have any leverage but just follow the developmentalist logic: money for life." Obviously, the local governments had little say when pressured into action by the provincial authority. Indeed, some argue that the whole campaign demonstrates a problem in the "institutional structure"[24] of command. As mentioned in the previous section, the idea of *fengshui* is so strong that rumors circulated suggesting that the grave-removal campaign was meant to destroy the *fengshui* of China's next leader.

Taking into account the resilient popular beliefs in *fengshui*, it is no wonder that graves are environmentally protected sites. A grave with auspicious *fengshui* should synthesize a microcosmic, well-balanced environment, which allows for the proper flow of wind and water, constantly passing through trees, soil, and grass. Annual ritual visits to graveyards during the Spring Festival and the Qingming Festival, or "tomb sweepings" (*saomu* 掃墓), involve pruning local vegetation and fixing terraces. In this way, graves represent not only the cosmological health of the dead, but the prosperity of the living based on the proper environmental maintenance.

However, the post-1949 state suggests that graves block deliverance to social and economic utopia in assuming that traditional burial practices will inevitably encroach upon arable lands. The grave-removal is part of a coherent linear progression of social projects toward the "utopian" socialist ideal as understood by the new regime. The first grave-removal campaign took place during the Great Leap Forward. Subsequent decrees released in the 1960s spelled out principles for the gradual introduction of cremation. The *Temporary Regulation for Mortuary Affairs*, passed by the State Council in February 1985, required a "gradual shift to cremation where the population is dense, arable land is scarce, and transportation is developed. Earthen burial is allowed elsewhere but should also aim towards cremation." On July 21, 1997, the State Council made this regulation permanent with an article forbidding burial in the following spaces: arable land, forests; urban parks, resorts, and sites of ancient relics; reservoirs, riverbanks, and their vicinity; catchments of water sources; and the sides of railways and highways. This article was taken quite literally in its provincial application in Henan via the *Henan Administrative Measures on Mortuary Affairs*, passed in 1999.

In Henan, one immediately finds that the forbidden areas overlap with all kinds of ideal *fengshui* sites, except for the mountains. Zhoukou and Nanyang, the other prefecture in focus during the 2012 campaign, are both topographically flat. Regulations forbidding burial by bodies of water or roads left the peasants with little alternative to burying on arable land. Private reserved land (*ziliudi* 自留地) for household use was often the best option owing to its closeness to one's house and ancestors' graves.

The interviews I conducted in 2013 demonstrated that ordinary villagers were not as irked by the campaign as elite reporters or academics were. Common people would have rather removed tomb-tops than send them to public columbaria for fear that, "the ancestors would be separated." In sending ash-urns to public columbaria, ancestors from a single family would risk being scattered in different shrines, thereby significantly increasing the chance of accruing bad luck from inauspicious *fengshui*. This belief is actually counter to the ancestral graveyard, where ancestors remain in the same vicinity and share the same *fengshui* landscape, a standard practice of earth burial.

This seemingly simple comment turned out to be very complicated, so I asked for further explanation. Firstly, the idea that ancestors should stay together reflects a deep *fengshui* conviction—a family graveyard should remain stable. A villager in Zhoukou said,

> For many years, we thought about moving them [graves] to better places. The year before the campaign, when I asked my uncle if we should do so, he replied over the phone that "we should think twice if we really want to do so. No one moves their ancestral graves unless something [bad] has happened. And, how can you be sure that the new place is better?"

Such interviews make clear the belief that only as a result of ill-fortune could a family think about relocating ancestral graves. "Moving without obvious reason may bring bad luck to us," another interviewee said.

Another salient apprehension—and probably the most important of all—was that the living would be separated from the dead, making it difficult to fulfill ritual obligations. Therefore, many villagers would rather remove tomb-tops themselves, before considering relocating the graves, and certainly before considering cremation. By removing tomb-tops, coffins could remain in place, albeit unmarked.

The villagers' opinions to this effect were reflected in media reports at the time of the campaign. For example, one villager in Zhoukou told a reporter that "those who oppose the policy and make trouble are outsiders, rich men with connections."[25] A villager from the Yang Village said,

> It's fine to level the ancestors' graves, but not the tombs of my father and mother. How can children dig up their parents' bodies after burying them? The graves have been in the field for generations. During the Spring Festival, families get together to get down on their knees and burn paper money before the graves. It's especially so now since young people work outside the villages most of the time. The only time they gather together is the annual offering to the ancestors. Otherwise, a family can't last.[26]

Similarly, one of my interviewees, Mr. Du from Zhoukou, told me that he did not mind leveling tomb-tops, but opposed the idea of moving his parents' and grandparents' graves. His reasoning was straightforward: tomb-tops could be rebuilt wherever and whenever when the campaign came to a close. Indeed, such was the case during the Spring Festival of 2013,

> To put it mildly, it [re-topping graves] is like the bamboo shoots after the rain. I mean, on New Year's Eve, many families pile the tomb-tops up after visiting the graves and offering incenses. Many new tomb-tops are even bigger than before, like new graves.

Re-topping tombs, or "rounding graves," has long been common practice for maintaining the *fengshui* of graves. It was widely practiced in Zhoukou for graves whose tops had been leveled. In a report in *China Youth Daily*, the primary concern was not about removing tomb-tops but finding peace for the ancestors. On New Year's Eve, one family was reported to have "rounded up" 166 graves they had leveled the previous year.[27] Another cadre, after removing his parents' tomb-tops, held an elaborate ceremony to re-top them. He repented and placed elaborate offerings by the grave, begging for his parents' forgiveness. He said that there had been forty students in his lineage admitted to universities since 1977, the year China reinstated the national college-entrance exam. The number exceeded all university offerings from nearby villages combined together. People believed it to be a sign of the wonderful *fengshui* of his ancestral graveyard, which had miraculously survived a flood that swept the village half a century earlier.

In a time of constant encroachment of industrial and real estate development, *fengshui* has become an effective tool for landscape maintenance and environmental protection throughout China (Zhu 2015). In an ethnographic study on the "spatial struggles" in Wenzhou, East China, Mayfair Yang elaborates on the importance of *fengshui* in the production of space. For earth burials, *fengshui* is about creating a friendly environment for the constant flow of cosmic forces. According to Yang, banning earth burials in Wenzhou seriously threatens the cosmic health of the people. Her analysis also applies to Henan villagers in the grave-removal campaign.

> The banning of earth burials in this rural area was a serious loss for the local people. Through fengshui siting of the graves of the dead, they provided a good afterlife for the deceased and ensured that contented ancestors would protect them and their descendants. Since tomb building involves fengshui expertise, the ban on earth burials narrowed an important arena for a fengshui production of space, where human corporeality is positioned in harmony with the pulses of larger cosmic forces running through the cosmos and the veins and contours of the earth. This archaic practice of aligning bodies, families, and lineages with space found its most important expression in the siting of burials. (Yang 2004: 731–732)

While most criticism of the campaign emerged from elitist concerns about land appropriation, provincial economic policy, and government accountability, *fengshui* remained the silent but ever-present concern of the villagers. *Fengshui* involves the synthesis of the terrace, trees, plants, rivers or lakes, and roads, which ensure proper flows of "wind" and "water." It stabilizes the landscape with the obligation of regular maintenance of soil and vegetation. Never an institutionalized religion and lacking any true power to balance the state, it is perhaps the intangibility of *fengshui* that lends it power as what James C. Scott (1987) calls the "weapons of the weak." For the villagers who resisted the campaign with relative success, the stability of a gravesite and its carefully selected *fengshui* features mattered more than its above-ground edifices. Hundreds of thousands of "rounded-up" graves tellingly demonstrate the deep and often invisible power of *fengshui*.

III. NEO-CONFUCIANISM: A REMEDY
TO STATE DEVELOPMENTALISM

Amid the public outcries about the political, legal, and economic aspects of the campaign, a group of self-identified Confucianists leveled criticism based on a new interpretation of Confucianism, "Neo-Confucianism" (*xinru jia* 新儒家)[28]. At the turn of the twenty-first century, Confucianism has experienced a strong revival. This revival has rested, in part, on externalizing China from the so-called West. The major players are the "mainland Neo-Confucianists" who are keen to uplift Confucianism into a state-recognized religion and shift the country to Confucian constitutionalism. Their attack on the campaign was an excellent chance to promote their version of Confucianism as an alternative to the state's utopia-cum-developmentalist policy.[29] On November 8, 2012, the Neo-Confucianists released a jointly signed appeal letter against the campaign, calling for its immediate abortion. This appeal came just six days before the State Council's Amendment to Article 20 of the *Regulations on Mortuary Practices*, which deprived Henan and other local governments of their power to remove graves. On November 21, the letter was updated to account for the Amendment. Altogether, about 400 self-identified Neo-Confucianists signed.[30]

The signers see Confucianism as an ideal remedy to China's "moral decadence." They generally believe that Confucianism can overcome the crisis of Western modernity by offering alternative social and bodily techniques relating to inner peace, communal solidarity, civility, and self-constraint (Tu 2010). In this way, Confucianism offers remedies to political corruption, economic disparity, social instability, and ecological disaster. Revivalists believe that modernity, characterized by liberal democracy and global capitalism, is no longer desirable for China. Instead, Confucianism will teach China and the world about the spiritual, moral, and ritual life, a sort of "civil religion" (Jensen 1997: 4). Many Confucianists promote the study of the Confucian classics in educational institutions and offer training to entrepreneurs, politicians, professionals, and spiritual seekers. Tellingly, the letter was double-dated using the Gregorian calendar (November 8, 2012) and the Confucian calendar (the 25th of the ninth month, year 2653 from the birth of Confucius). The letter was even officially released by four Confucian websites: *Chinese Confucian Religion* (*zhonghua rujiao wang*), *Confucian China* (*rujia zhongguo wang*), *Confucius 2000* (*kongzi 2000wang*), and *China Mainstream Culture* (*zhonghua zhuliu wenhua wang*). Indeed, just a few months before signing the appeal letter, the leading figure of the Neo-Confucian movement, Jiang Qing published the famous "A Confucian Constitution for China" in the *New York Times*, together with a Western supporter of Confucianism, Professor Daniel Bell at Shandong University.[31]

A week after the appeal's publicization, the lead drafter, Professor Yao Zhongqiu of Beihang University even hosted a symposium on "the Religious, Cultural and Legal Gaze of the Grave-removal Campaign."[32] The participants of the symposium theorized on the reasons behind the grave-removal campaign. Unlike mainstream media, which focused on economic development, land grabbing, or unequal economic measures, the Neo-Confucianists determined the root of the problem to lie in the state's "overly Western" developmentalist outlook.

Xu Xin, Gan Chunsong, and Wang Zhiping all argued that if there were a Chinese religion, it should be ancestor worship. The organizer of the symposium, Professor Yao Zhongqiu, explicitly and daringly charged that the contemporary Chinese "power mechanism" is actually a blueprint for foreign actions against Chinese civilization.

In the symposium and the appeal letters, Neo-Confucianists dichotomized the local government as a Westernized regime with a developmentalist mindset that would "damage Chinese culture." Radical developmentalism would "hurt the feeling of the masses," because even "an ethical Confucian king of the past would protect graves." The letter called for

(1) an immediate ending of the grave-removal campaign;
(2) the central government's intervention: "[it should] understand that ancestor worship is the basic belief of the citizens, and should respect the people's freedom of this belief"; and
(3) public action: "the wise public and media should . . . maintain the freedom of belief and the customs of the Chinese people, cherish Chinese historical and cultural traditions, and promote the reconstruction of 'cultural China' and 'Chinese musical protocol.' "[33]

Some of the narratives constructed by the Neo-Confucianists in their argument contrasted the "religious beliefs of the Chinese People" with the radical developmentalism of the state. By dichotomizing the "radical transcendence of Western values" with "the harmony of Confucianism," they relied on the presumed existence of an organic, lived environment in which *fengshui* conceptions were honored, thus offering a remedy to environmental degradation.

Establishing "ancestor worship" as "the belief of the Chinese People" was a crucial step for the self-identified Neo-Confucianists in their endeavor to condemn the grave-removal campaign. Belief in ancestor worship relieves anxiety about the processes of life and death by giving ephemeral life an eternal meaning. By depicting the campaign as "barbaric conduct that seriously violates the freedom of belief, damages Chinese culture, and hurts the feeling of the masses," the Neo-Confucianists accused the Henan provincial and prefectural governments of damaging "ancestor worship, the most important belief (xinyang 信仰) of the Chinese people." The appeal letter quoted ancient classics on Confucian cosmology, with phrases such as, "Heaven creates living things; ancestors create humans":[34]

> Since ancient times, Chinese people have firmly believed in the merits and virtues of ancestors, and upheld the idea of "celebrating one's origin by revering Heaven and following the ancestral ways." As such, the living must bury their deceased relatives with proper rites, seal their graves, and offer sacrifices throughout the four seasons. There is a classical saying which goes, "When alive, serve with rites; when deceased, bury with rites and make ritual offerings." Chinese also believe that they will happily join their ancestors in another world after death. Descendants thus provide offerings for future meetings with them as shown in another classical saying, "Make offerings as if [the ancestors are] present; offer to the gods as if they are present." As ancestors' burial sites, graves are important material abodes of belief. However, the grave-removal campaign in Henan Province ignored this thousands-year-long, widespread belief. It seriously violated and deprived the citizens' rights of freedom of belief, humiliating both men and gods, and shattering law and reason.

Organized actions such as appeal letters against a provincial government were rare. However, this action on the part of the Neo-Confucianists was successful partly because it subtly played on the concept of "religion." Their arguments referred to constitutional religious freedom without an appeal to "religious pluralism," and ignored religious diversity in Henan. Strikingly, the scholars were able to deploy Confucianism even though it is not one of the five legal religions of contemporary China. While the Constitution states that Chinese citizens enjoy the right of "freedom of *religious* belief," the letter used the term "freedom of belief" and omitted the word "religious." The Neo-Confucian scholar-activists behind the letters sublimated ancestor worship as "the most innocent, widespread, dear, and powerful natural belief of the Chinese people," in contrast to the constitutional right for "freedom of religious belief," which applies only to state-recognized and institutionalized religions such as Buddhism, Daoism, Christianity, Catholicism, and Islam. The strong religious tone of their writings was thus relegated to a description of the transcendental world—heaven, dead ancestors, and gods, which may all turn indignant against the wrong-doing of the government and humans. It was no small matter that two of the letter drafters were also law professors who resorted to highlighting a legally non-provident clause on "freedom of belief" rather than invoking Land Administration Law or Property Law.

Appealing to the government to deliver a constitutional right was a bold action. However, the action did not suggest religious pluralism, but religious singularity. In the letter, scholars often used the expression "the Chinese people" (*zhongguo ren* 中國人), a term interchangeable with "citizens," presuming that all Chinese citizens are followers of the Confucian classics. The letter's imagining of the nation as a community of homogenous people is explicit; creating a regime of authenticity by which all Chinese people in Henan Province were inherently Confucians who venerated their ancestral tombs.

However, as stated earlier, not only is Confucianism not a legally recognized religion, it is far from being a belief shared by all Henan people. Henan is in fact one of the few provinces densely populated by Muslims as well as Christians, both of whose religions are legally recognized. This is especially the case in Zhoukou and Nanyang. During the campaign, policy documents explicitly stated that Muslims graves were to be exempted for fear of protest. However, the exemption was not given on the basis of religious minority status, but on the grounds of respecting the "particular customs" of the Hui *ethnicity*. Other ethnicities who live in the province did not enjoy this positive discrimination policy. More interestingly, it seems Muslims were not considered citizens under the government's vow to have every new rural death end in cremation. Muslim populations also remain largely silent. They seldom produced an opinion on the polyphonic controversy on the campaign.

While Muslim graves were not destroyed, Christian graves were. Videos on different websites show graves topped with crosses being destroyed. In Zhoukou Prefecture alone, Christianity has spread to almost every single township, mostly through "house churches." Generally, Chinese Christians respect the deceased but refuse to participate in "idolatrous" traditional Chinese funeral rites. Therefore, when a Christian dies, the funeral is usually marked with solemnity, soberness, and simplicity, and lacks rituals (Lim 2013). The Christian graves are more a memorial than of a site of worship. Given these differences, some Christians supported the grave-removal campaign. Many house churches adhere to a radical theology that despises physical representations such as bones and graves. One house church leader stated that Christians deemed the campaign "civilized" (*wenming* 文明). "We believe there are only bones in the tombs, not souls. Please start the campaign with our [ancestral] graves."[35] The statement is often quoted by the state authority, although this same authority often denied the existence of house churches, because of their refusal to join the Christian, state-monitored Three-Self Patriotic Movement Church.

In their condemnation of the government, the Neo-Confucianists some-times also levied criticism at the Chinese state for blindly following a devel-opmentalism deeply embedded in Western-style mindsets. Han Xing called for rebuilding beliefs, suggesting:

> The centrality of politics leads to the inertia of the hegemonic control of soci-ety, leading to personal idolatry, philosophy of struggle, and the mindset of the Cultural Revolution. For a long time, we have been promoting materialism and atheism against any religion, destroying the traditional belief system of the Chinese people. These include the principle of the Heavenly Way (天道); the Three Teachings of Confucianism, Daoism, and Buddhism Integrated into One (三教合一); the relationships of Heaven, Earth, Emperor, Parents, and Teach-ers. [This destructiveness of tradition] is very frightening.

Chen Bisheng, another outspoken Neo-Confucianist, criticized the campaign of sabotaging customs, "The mindset of the local government about their rule is a revolutionary one like 'Destroying the Four Olds.' It adopts a radi-cal materialist attitude, destroying traditional Chinese customs without any humaneness and exterminating the traditional lifestyle of the Chinese peas-ants without any mercy."

The Confucianists who produced the open letter might be easily regarded as a civil society organization that successfully changed the state's mind, given the fact that it was celebrated as one of the top ten events in "civil society" of 2012 by the Civil Society Research Centre at Peking University. However, this is a rather-illusive image. As Cao Nanlai contends, the "state-society dichotomy" reifies a domination-resistance model that "often reflects the researchers' concern for moral clarity and an old political logic rather than the views of the local believers" (2011: 7). Indeed, the self-identified Confucianists did not conceive of themselves as being against the state. On the contrary, the Confucian activism consciously attempted to remove China from the West. Thus, the true dichotomy at work was not one of the state versus society, but one of China versus the West.

Dichotomizing the radical developmentalism of the state and contrasting it with the spontaneous beliefs of "the Chinese people" manifests the Neo-Confucian conviction of the China-West dichotomy. This dichotomy is evi-dent in some key self-identified Confucian scholars, such as Jiang Qing, Han Xing, Kang Xiaoguang, and Yao Zhongqiu, who all signed the appeal letters. As discussed in this section, part of the narrative constructed by the Neo-Confucianists in their argument against the campaign contrasted the religious beliefs of the Chinese people with the radical developmentalism of the state.

There are many quotations from the Confucian classics demonstrating the ethics of environmental protection. For example, in one of the Confucian classics—the *Book of Rites* 《禮記》, there is a quotation from Confucius about the proper correspondence between filial piety and sacrifice, "When cutting down a tree or slaughtering an animal beyond its time, he violates the principle of filial piety." Such excerpts suggest that one should not terminate a living being without respecting its life cycle. Unnecessary or premature killing is thus considered to be like being disrespectful to one's own parents. In a similar vein, the protest against the grave-removal campaign can be understood as part of the contemporary Neo-Confucian environmental movement. Indeed, environmentalists have often attempted to reorient the Confucian anthropocentric outlook to better highlight ideas of harmony with nature, yearning for longevity, and sympathy to hardship, as they relate to sustainability. The International Confucian Ecological Alliance (ICEA) tries to combine Confucian wisdom and ecological science to foster a global network to raise awareness of the world's severe ecological crisis. Today, Chinese scholars are also exploring ancient "ecological thought" embedded in the Confucian classics, in light of contemporary environmentalism (Chen 2012; Qiao 2013). These attempts are compelling, though lacking in necessary empirical studies of current events.

IV. CONCLUSION

In this chapter, I have argued that the Henan Grave-Removal Campaign was part of a state developmentalist agenda. By reclaiming at least some amount of arable land through removing rural graves, the Henan authorities hoped to create more industrial land. However, this industrial land would inevitably be allotted to high-pollution industries. The public outcry played a vital role in finally leading the central government to intervene and call off the campaign. However, the villagers themselves were well-prepared. Marshaling "weapons of the weak," they removed tomb-tops instead of relocating urns and ashes. The unspoken reason for the villagers' relative acceptance of the policy was concern for maintaining stable geomancies synthesizing the elements of land, water, trees, and roads. In other words, there was a widespread popular conviction that cosmic forces would be damaged if the state were to overdevelop the environment.

Among the most prominent elitist condemnations of the campaign was the Neo-Confucianist narrative of ancestor worship as a Chinese religion held dear by a homogeneous "Chinese people." By contrasting ancestor worship with the radical developmentalism of the state, the Neo-Confucianists dismissed the state power as Westernized. Ancestor worship, on the other hand, embodied values of harmony between humans and nature and was thus presented as fostering a milder, sustainable development. Their proposed dichotomy of China versus the West created hope for what Prasenjit Duara argues is an alternative to the crisis of Western modernity. Interestingly, the elitist construction of ancestor worship coincided with the villagers' silent, geomantically-informed resistance. The values of both the villagers and the Neo-Confucianists can be seen as quasi-religious—given their lack of institutionalization—and de facto environment-friendly.

The backlash to the Henan campaign transcends the discussion on how Chinese religious ideas can offer de jure guidance for the environmental crisis in the epoch of Anthropocene. Chinese ancient literature is full of wise sayings such as that found in the *Huainan Zi* of the second century BCE, which states, "Do not fish by draining the pond; do not hunt by burning the wood. That is the decree of the late kings" (Liu 1989: 308). There is never a shortage of historical reference material when it comes to contemporary issues. However, reality often demands a powerful and effective social institution, without which ideas remain ideas and are never put into action.

This point brings us back to Prasenjit Duara's argument on hope. In his examination of grassroots activism across Asia, Duara (2015) convinces us that religious activists—Confucian, Buddhist, or Hindu—who continue a dialogue on their respective transcendent values stand to offer an alternative to Western hegemonic thought. Billioud and Thoraval (2015) have shown how popular Confucianists (*minjian rujia*) revive Confucian teaching through school curriculum, theological codification, Confucian festivals, and Confucian capitalization. This chapter has highlighted yet another venue for Confucianism, namely, the *non-institutional* practices of geomancy and ancestral veneration of graves. Though most of the locals who were reluctant to remove their parents' graves are not self-identified Confucianists, and would never think of participating in any form of organized protest, they were not necessarily irrelevant to the elite Confucian activists involved in the protests. In fact, many of these same elites were born and raised in Henan. In the end, institutionalized activism is not necessarily effective in China in that it is too readily identified and dismantled by the state. Any organization can be easily dissolved. The silent, inarticulate, non-cooperative, and "everyday forms" of resistance like those demonstrated by Henan peasants are perhaps more forceful. They are Confucianists without being Confucian and resist without being resistant. Besides grassroots activism as depicted by Duara, and Billioud and Thoraval, non-institutional activists may be the real "hope," if there is any at all.

Are these ideas good enough for religious environmentalism? Probably not if we are hoping for an emerging institutionalizing power to remedy the global environmental crisis. However, we cannot accept fatalism in the face of the opened Pandora's box that releases everything destructive to humanity. The environmental crisis will persist as long as the world encourages fierce productivism and nationalistic competition. Technologies for policing—big data, artificial intelligence, 5G networks, facial recognition, citizen credit—are advancing with geometric speed. Even "indigenous wisdom" itself has to rely on the inverse Orientalism of the dichotomy of China versus the West.

Human beings have committed numerous atrocities in the course of our evolution, violating many essential laws and acting against the best interests of our survival as no other species has done. We waste food, we exploit our peers, we legitimate selfishness, and we create "bullshit jobs" (Graeber and Cerutti 2018). Society is organized according to positive feedback not founded in the biological realm, except for terminal diseases such as cancer. As a biological species, we exist because we follow the negative feedback of automatically canceling out any emerging mechanism. The positive feedback mechanism of the society is persistently against evolution to the point that the significance of the human is no longer biological, but geological—we turn the globe against ourselves. Faced with the unprecedented disaster of the COVID-19 pandemic, and struggling to restore the "normal" life, we dream of flying to any corner of the world to meet anyone at will, at little cost—a true reflection of how greedy humans have become. Perhaps *Homo sapiens*, or better yet, *Homo stultus*, was probably created with a self-destruct mode from the very start.

NOTES

1. Research for the present work was conducted through a small project while teaching *Sociology of Religion* in China, 2014. Follow-up interviews were conducted at the end of the same year. The author would like to thank project participants for their efforts in fieldwork and for many inspiring discussions. In particular, thanks go to Sun Guangyang, Sun Chen, Xiang Yingqian, Huang Zhengzheng, Li Xiaoxuan, and Cao Yuze. Nate Sims and Mayfair Yang contributed to the editing and polishing of this chapter.

2. "Among Ten Publicized Civil Society Events of 2012, Environmental Mass Incidents Rank First" (2012 年十大公民事件出爐 多地環保群體事件居首), Jan 28, 2013, http://news.sohu.com/20130128/n364786766.html. Accessed on April 20, 2014. The ranking is conducted by Centre for Civil Societies at Peking University.

3. 1 *mu* =666.7 square meters.

4. Commune of the People's Government of Zhoukou Municipality (周口市人民政府政府公報), "Speech by Comrade Yue Wenhai Delivered at the Municipal CCP Committee Meeting for Rural Work and Poverty Alleviation" (岳文海同志在市委農村工作暨扶貧開發工作會議上的講話), http://www.zhoukou.gov.cn/html/4028 81121bfdf857011bfe2613590038/2012032808073907.html, Feb 29, 2012. Accessed on April 25, 2014.

5. Meng, Xiangchao (孟祥超). "Investigating Grave-removal for Re-cultivation" (河南周口試點平墳複耕調查), July 18, 2012, *The Beijing News* 《新京報》, http://www.bjnews.com.cn/news/2012/07/18/210958.html. Accessed on May 6, 2014.

6. Liang, Weifa (梁為發). "The Land Game Behind the Zhoukou Grave-Removal" (周口平墳背後的土地博弈), *Metropolitan Hebei* 《燕趙都市網》, January 2, 2013, http://hn.ifeng.com/zixun/yaowen/detail_2013_01/03/514473_0. html. Accessed on May 6, 2014.

7. She, Yiwei (余以為). "Grave-Removal Campaign Caused by the Red-line of 18 million mu", ("平墳運動禍起十八億畝紅線"), Nov 23, 2012, http://blog.caijing. com.cn/expert_article-151516-44076.html. Accessed on May 6, 2014.

8. Law Web (法制網). "Media Uncovered Land Economy Behind Henan Grave-Removal: Government Sacks Billions" (媒體揭河南平墳背後土地賬：政府可獲利百億元), Dec 12, 2012, *Sina News Center* 《新浪新聞中心》, http://news.sina.com.cn/c/sd/2012-12-18/113025838451 .html. Accessed on April 20, 2014.

9. "Henan 2012 Environmental Release finds Kaifeng Zhengzhou at Bottom in Air Quality", (河南公佈 2012 環境狀況開封鄭州空氣品質倒數), June 5, 2013, http://hn.ifeng.com/zixun/redianguanzhu/detail_2013_06/05/871701_0.html. Accessed on August 10, 2019.

10. Item "Huo Daishan" (霍岱珊) https://baike.baidu.com/item/%E9%9C%8D %E5%B2%B1%E7%8F%8A. Accessed on May 7, 2014.

11. "China Central Television Gets Ingenious Reply While Investigating Wheat Pollution Case" (央視採訪污染小麥遇神回答), March 29, 2013, http://news.sohu .com/20130329/n370839606.html. Accessed on August 10, 2019.

12. "704 Dead in 15 years in a Polluted Henan 'Cancer Village': Deputy Mayor under Inspection" (河南因污染現癌症村15年死704人 副市長被責檢查), *China Web* (中國網), Sept 26, 2007, http://news.china.com.cn/txt/2007-09/26/content_89 57484_2.html. Accessed on August 10, 2019.

13. See Note 8.

14. "Secretary Lu Zhangong Gives Important Comments on the Work of Zhoukou Grave-Removal and Reclamation" (盧展工書記對周口平墳擴耕工作做出重要批示), August 4, 2012, https://web.archive.org/web/20130531002713/. http://news .hexun.com/2012-08-10/144633385.html. Accessed on April 20, 2014.

15. See Note 8.

16. Ibid.

17. Yu, Mengjiang(于夢江). "Feeder Changed Message on 'Chuji Ancestral Graves'" ('副處級祖墳'爆料者改口"), May 8, 2012, *Guangzhou Daily Net* 《廣州日報網》, http://gzdaily.dayoo.com/html/2012-05/08/content_1694930.html. Accessed on April 20, 2014.

18. "Legislator Zhao Keluo against Grave-removal: Couldn't Keep Silent against Conscience" (反平墳政協常委趙克羅：違背良心的事不能再保持沉默), *Phoenix i-Talk* 《鳳凰自由談》, Jan 22, 2013, http://news.ifeng.com/exclusive/fangtan/speci al/zhaokeluo/. Accessed on April 27, 2014.

19. "How Ancestral Graves of the Officials Will become 'Graves of Famous People'" (有多少官員祖墳會變成'知名人士墓'), May 8, 2012, *Xinhua Net*《新華網》, http://news.xinhuanet.com/comments/2012-05/08/c_111904695.html. Accessed on April 20, 2014.

20. Meng, Xiangchao (孟祥超). "Two Senior Party Members in Zhoukou Killed in Accident by Tombstones They Toppled" (河南周口老黨員帶頭平墳挖墓碑不慎將兩人砸死), *Tencent Net*《騰訊網》, Nov 23, 2012, http://news.qq.com/a/20121123/000070.html. Accessed on April 11, 2014.

21. "Decision of the State Council to Amend and Repeal Certain Administrative Regulations (No. 628)" (國務院關於修改和廢止部分行政法規的決定 [第628號]), Nov 16, 2012, *The Central People's Government of the People's Republic of China* (中華人民共和國中央人民政府網站), http://www.gov.cn/flfg/2012-11/16/content_2269504.html. Accessed on April 11, 2014.

22. "Ministry of Agriculture Spokesman on Henan Grave-Removal: Goodwill with Bad Measures" (農業部發言人談河南平墳：願望好但辦法欠妥), Dec 25, 2012, *International Online* (國際在線), http://news.qq.com/a/20121225/001021.html. Accessed on April 22, 2014.

23. Zhao, Fasheng (趙法生). "Crime and Punishment of Henan Grave-Removal Officials" (河南平墳官員的罪與罰), March 12, 2013, *Aisixiang*《愛思想》, http://www.aisixiang.com/data/61993.html. Accessed in April 22, 2014. See also Wei, Dedong (魏德東). "Cherish People's Sacred Pursuit" (呵護人民的神聖追求), June 18, 2014, http://news.ifeng.com/a/20140618/40783689_0.html. Accessed on August 9, 2019.

24. Kong, Deji (孔德繼). "Who did the Ghosts Offend?: Dynamics and Institutional Fault in the Zhoukou Grave-Removal" (死鬼惹了誰? —— 論周口平墳的動力來源與制度癥結), Nov 12, 2012, *Netease Microblog*《網易博客》, http://condj.blog.163.com/blog/static/953428272012101233297 49/. Accessed on April 23, 2014.

25. *China Youth Daily* (北京青年報). "Zhoukou: Village Party Chief Level First, Opposers Are Rich" (周口：村支書帶頭平墳 反對平墳是有錢的有人的), Feb 25, 2014, *Observer* (觀察者), http://www.guancha.cn/society/2013_02_25_128233.html. Accessed on April 20, 2014.

26. "Investigation on Zhoukou Grave-Removal: Peasants Still Resist" (周口平墳調查：有農民仍抵抗), Dec 15, 2012, *South Reviews* (南風窗), http://style.sina.com.cn/news/p/2012-12-15/0751111713.html. Accessed on April 20, 2014.

27. Guo, Jianguang (郭建光). "Zhoukou Grave Rounding-up: Souls at Peace and Hearts at Ease" (周口圓墳：魂靈安息人心安放), Feb 27, 2013, *China Youth Online* (中青在線), http://zqb.cyol.com/html/2013-02/27/nw.D110000zgqnb_20130227_3-09.html. Accessed on April 20, 2014.

28. The beliefs of the self-identified Neo-Confucianists of contemporary China differ from Neo-Confucian beliefs as they existed during the heyday of late imperial China. Here within, Neo-Confucianists mainly refer to Confucianists in post-Mao, mainland China.

29. Peng, Yongjie (彭永捷), "Each of us is a Neo-Confucianist" (我們每個人都是新儒家), Jan 20, 2013, *People's Daily Online*《人民網》, http://theory.people

.com.cn/n/2013/0204/c112851-20427366.html. Accessed on April 20, 2014. These self-identified Neo-Confucianists include people like Chen Ming (陳明), Shi Pu (史璞), Ren Feng (任鋒), Zheng Fengtian (鄭風田), and Yao Zhongqiu (姚中秋).

30. "Expedited Appeal Letter by All Walks of Life for Immediate Stop of the 'Grave-removal Campaign" (《社會各界關於立即停止"平墳運動"的緊急呼籲書》), November 8, 2012, and "Second Expediated Appeal Letter by All Walks of Life for Immediate Stop of the 'Henan Grave-removal Campaign" (《社會各界人士關於河南平墳運動的第二次緊急呼籲》), Nov 21, 2012, https://www.rujiazg.com/article/3085. Accessed on April 20, 2014.

31. Jiang, Qing and Daniel Bell. "A Confucian Constitution for China," July 10, 2012, *New York Times*, https://www.nytimes.com/2012/07/11/opinion/a-confucian-constitution-in-china.html. Aaccessed on December 12, 2016.

32. See Yao Zhongqiu's Sina Microblog, Nov 28, 2012, http://blog.sina.com.cn/s/blog_543973110101a6u4.html. Accessed on April 21, 2014.

33. "Chinese Music Protocol" (禮樂中國) was a ritual institution in ancient China. The ancient Confucians generally held that only a proper musical institution could manifests the ethics of the rites.

34. See note 30 of this chapter.

35. See "Tomb-removal Encourages Foreign Religions," https://www.zgxcfx.com/Article/51851.html. Accessed on April 21, 2014.

ENGLISH AND WESTERN LANGUAGES

Billioud, Sebastien and Joel Thoraval. 2015. *The Sage and the People: The Confucian Revival in China*. Oxford: Oxford University Press.

Bruun, Ole. 2003. *Fengshui in China: Geomantic Divination between State Orthodoxy and Popular Religion*. Copenhagen: NIAS Press.

Cao, Nanlai. 2011. *Constructing China's Jerusalem: Christians, Power and Place in Contemporary Wenzhou*. Stanford, CA: Stanford University Press.

Chau, Adam Yuet. 2019. *Religion in China: Ties that Bind*. Medford, MA: Polity.

Descola, Philippe. 2013. *Beyond Nature and Culture*. University of Chicago Press.

Duara, Prasenjit. 2015. *The Crisis of Global Modernity*. Cambridge: Cambridge University Press.

Fei, Hsiao-t'ung, ed. 1992 [1948]. *From the Soil: The Foundations of Chinese Society*. Translated by Gary G. Hamilton, and Zheng Wang. Berkeley: University of California Press.

Feuchtwang, Stephan. 1974. *An Anthropological Analysis of Chinese Geomancy*. Vientiane: Vithagna.

Girardot, Norman J., James Miller and Liu Xiaogan, eds. 2001. *Daoism and Ecology: Ways Within a Cosmic Landscape*. Cambridge, MA: Harvard University Press.

Graeber, David and Albertine Cerutti. 2018. *Bullshit Jobs*. New York: Simon & Schuster.

Hsu, Francis. 1948. *Under the Ancestors' Shadow*. New York: Columbia University Press.

Jensen, Lionel M. 1997. *Manufacturing Confucianism: Chinese Traditions and Universal Civilization*. Durham: Duke University Press.

Lim, Francis, ed. 2013. *Christianity in Contemporary China: Socio-cultural Perspectives*. New York: Routledge.

Mueggler, Erik. 2001. *The Age of Wild Ghosts: Memory, Violence, and Place in Southwest China*. Berkeley: University of California Press.

Scott, James C. 1987. *Weapons of the Weak: Everyday Forms of Peasant Resistance*. New Haven: Yale University Press.

Taylor, Charles. 2007. *A Secular Age*. Cambridge: Belknap Press of Harvard University Press.

Tu, Weiming. 2010. *The Global Significance of Concrete Humanity: Essays on the Confucian Discourse in Cultural China*. New Delhi: Center for Studies in Civilizations and Munshiram Manoharlal Publishers.

Whyte, Martin. 1988. "Death in the People's Republic of China." In *Death Ritual in Late Imperial and Modern China*, 289–316.

Yang, Mayfair. 2004. "Spatial Struggles: Postcolonial Complex, State Disenchantment, and Popular Reappropriation of Space in Rural Southeast China." *The Journal of Asian Studies*, vol. 63, no. 3, p. 719–755.

Yang, Mayfair. 2008. "Introduction." In *Chinese Religiosities: Afflictions of Modernity and State Formation*, edited by Mayfair Yang. Berkeley: University of California Press.

Zhu, Xiaoyang. 2015. *Topography of Politics in Rural China: The Story of Xiaocun*. Hackensack, Singapore: World Scientific Pub. Co.

CHINESE LANGUAGE

Chen, Yexin 陳業新. 2012. 《儒家生態意識與中國古代環境保護研究》(*Studies on Confucian Ecological Awareness and Chinese Ancient Environmentalism*). Shanghai: Shanghai Jiaoda Chubanshe.

Liu, Wendian 劉文典. annotated.1989. 《淮南鴻烈集解》(Collected Annotation for *Huinan Honglie*). Beijing: Zhonghua Shuju.

Liu, Ziqian, and Chen Tao 劉子倩, 陳濤. 2012. "The Zhoukou Tomb-removal Storm" (周口平墳風暴). *China Newsweek* (《中國新聞週刊》), no. 44, pp. 40–42.

Qiao, Qingju 喬清舉. 2013.《儒家生態思想通論》(*Introduction to Confucianist Ecological Thought*). Beijing: Beijing Daxue Chubanshe.

III

SENTIENT BEINGS

ENGAGING WITH ANIMALS AND DIVINITIES IN DREAMS AND RITUALS

Chapter 5

The Non-Anthropocentricity of Dreaming in Late Classical and Medieval China*

Robert Ford Campany

> You and I are both things. What nonsense, that one
> of us should think it is the other which is the thing!
> 且也若與予也皆物也，奈何哉其相物也！
> A certain old oak tree, addressing a carpenter[1]

It could be argued that anthropocentrism starts, conceptually and metaphorically at least, with models of the human person. Not every model or metaphor for the human being would involve placing ourselves and our "cultures" at the center of reality, starkly juxtaposed against the nonhuman, the merely "natural." Jettisoning anthropocentrism, or at least loosening its grip on us, might begin with imagining different models of the sorts of beings we are and different possibilities for relationships with extra-human beings. For this, works of imagination produced by people of other times and places can be useful.

In this chapter, I sketch a few examples of the different sorts of models of ourselves and of our relations with other beings on which we might reflect. They are drawn from the extensive premodern Chinese discourse on dreams. All of us humans, it seems, experience dreaming—dogs, cats, horses, and other creatures do too—but what happens when we dream, who the "I" is who experiences dreaming, and what can or should be done in response has been imagined in many ways.

---

* Portions of this essay previously appeared in: Robert Ford Campany, *The Chinese Dreamscape 300 BCE – 800 CE*, Harvard-Yenching Institute Monographs 122, Harvard University Asia Center, Cambridge, MA and London, 2020.

I. THE LONELY MODERN DREAMER?

As far as I am aware, all of the important modern, Western models of dream-ing—important in the sense that they are the ones people often write about or are professionally initiated into, constituting a quasi-official or quasi-com-mon-sense view—presuppose that dreams are made by, and are fundamen-tally about, the dreamer. Freud: "It has been my experience that every dream without exception deals with oneself. Dreams are absolutely self-centered" (Freud 1999: 246). Jung: "Every interpretation of a dream is a psychologi-cal statement about certain of its contents," and the therapist arrives at this meaning for the dreamer "by a methodical questioning of the dreamer's own associations," a process Jung called "taking up the context": "This consists in making sure that every shade of meaning which each salient feature of the dream has for the dreamer is determined by the associations of the dreamer himself" (Jung 2010: 69–71). Gestalt psychologist Frederick Perls, speak-ing to a patient: "This is your dream. Every part is a part of yourself" (Perls 1969: 163). One major neuroscientific theory of the function of dreams sees them as products of the brain's nocturnal consolidation of memories—the constant building and rebuilding of private knowledge structures (Rock 2004: 77–100). Another understands dreams as primarily regulators of emo-tions (Rock 2004: 101–120). Whatever their differences, all these views see dreams as being about the dreamer, as being made by the dreamer's mind or brain, and as reflecting the dreamer's past experience or present concerns. Dreams are postcards from ourselves to ourselves.

I do not know when this presupposition gained its current prominence. Perhaps it was in the nineteenth century. In any case, there is a prevalent, powerful modern myth that lies ready to hand to frame what happened: the myth of disenchantment, based in part on Max Weber's *Entzauberung* thesis. Charles Taylor, who doesn't seem to view it as myth but rather as fact, sum-marizes some of its key elements:

> Everyone can agree that one of the big differences between us and our ancestors of 500 years ago is that they lived in an "enchanted" world and we do not. We might think of this as our having "lost" a number of beliefs and the practices they made possible. Essentially, we become modern by breaking out of "super-stition" and becoming more scientific and technological in our stance toward our world. But I want to accentuate something different. The "enchanted" world was one in which spirits and forces defined by their meanings (the kinds of forces possessed by love potions or relics) played a big role. But more, the enchanted world was one in which these forces could shape our lives, both psychical and physical. One of the big differences between our forerunners and us is that we live with a much firmer sense of the boundary between self and other. We are "buffered" selves. We have changed. (Taylor 2011: 38–39; Puett 2012: 109–113)

Meanwhile, palliatives have been offered to this disenchantment macro-narrative. Jane Bennett, to mention a single writer, has pointed to "survivals," as it were (not her term), of enchantment on this side of the supposed gulf between our era and all that went before (Bennett 2001). Her project is meant to comfort us in our arid modernity. Others, however, have rejected the disenchantment trope itself as a strange modern myth, a story we tell ourselves about ourselves that is massively belied by evidence. As Bruno Latour puts it, "Haven't we shed enough tears over the disenchantment of the world? Haven't we frightened ourselves enough with the poor European who is thrust into a cold soulless cosmos, wandering on an inert planet in a world devoid of meaning?" (Latour 1993: 115). Jason Josephson-Storm has compiled a big dossier of counterevidence, arguing that "these new philosophers [such as Bennett], like the poststructuralists they seek to replace, are rebelling against a hegemon that never achieved full mastery. The enchanted ontologies and spiritualized orientations to nature they describe as missing have been available all along" (Josephson-Storm 2017: 5). And sure enough, if we peer beneath the quasi-official, radically anthropocentric view of dreaming, it turns out that modern societies have a much more vibrant set of ideas about dreaming than the myth of disenchantment would suggest.

For example, one survey of attitudes toward dreams held by students from the United States, South Korea, and India found that a majority of subjects treat the content of their dreams as more meaningful than the content of waking thoughts. This same study noted that many respondents think some of their dreams provide insight into future events. Yet it is telling that the study's authors, unable to square this belief with their own assumptions, confidently wrote it off: "Horrible dreams about plane crashes might be evidence that someone is anxious about a meeting they are scheduled to attend, but such dreams are certainly not evidence that a plane crash is imminent" (More-wedge and Norton 2009: 261). (Dreams can't be taken as predicting future events, after all, if their only possible topic is dreamers' anxieties.) Over half the subjects surveyed in another study believe that the dead "can actually visit us in dreams" (Kunzendorf et al. 2007–2008), or (in yet another study) that dreams carry important messages (Mazandarani et al. 2018). Many other examples could be cited. Taken together, these survey-based inquiries, partial and under-interpretive though they are, show that the quasi-official modern view of dreaming does not begin to capture how modern populations actually think about and respond to their dreams.

We need more studies of "folk" theories of dreaming in Europe and North America both before and since the rise of the discipline of psychology.[2] Meanwhile, for modern dreamers seeking alternative models with which to (further) enchant their dream lives or enlarge their views of what their dreams are and what they could be, medieval China offers rich possibilities.

II. THE EXORCISTIC MODEL

If you were a literate person at any point in time from the Qin up until around 800 CE, and you woke from a nightmare, you might, if you had access to it, turn for help to a text prescribing a certain characteristic response. A closely similar procedure is prescribed in two daybooks (*rishu* 日書) written on bamboo strips, buried in 217 BCE and recovered at Shuihudi (睡虎地) in Hubei province in 1975, as well as in a Dunhuang manuscript (Pelliot 2682 recto) dating from the late Tang or slightly earlier. What this similarity entails is that over this longue durée there perdured a specific understanding of what nightmares are and how to "cure" them. The Shuihudi a daybook passage reads:

> When a person has foul dreams, on wakening then unbind the hair, sit facing the northwest and chant this prayer: "Heigh! I dare to declare you to Qinqi. So-and-so[3] has had foul dreams. Flee back home to the place of Qinqi. Qinqi, drink heartily, eat heartily. Grant so-and-so great broadcloth.[4] If not coins, then cloth. If not cocoons, then silkstuff." Then it will stop.[5]

To readers who have had a foul dream (*e meng* 惡夢), each text prescribes a ritual which includes unbinding the hair, facing in a particular direction, and chanting an invocation or prayer to a divine figure (here named Qinqi) who oversees the spirits responsible for causing foul dreams. The incantation urges this deity to enjoy the offerings made by the dreamer and, in return, to prevent further nightmares. The incantation also commands the dream-causing spirits to return to their overseer. It closes with the request that the divine overseer bestow material blessings in exchange for the offerings.

For this and similar texts—and, again, we have many more of them, strewn across the "religions" historians often imagine as immaculately distinct from one another (Campany 2003)—foul dreams are not artifacts of sleepers' brains. They are inter-being interactions. More specifically, they are assaults by exogenous beings who breach the bounds of the person left vulnerable by sleep. It's these beings who are the agents of the dream; the dreamer is merely a passive subject. The texts prescribe a relatively simple exorcistic ritual in response—a ritual in which the dreamer reclaims agency. But they do more than this. They render the dreamer's experience intelligible as resulting from a specific, frequent sort of interactive process: the dreamer has been assaulted, but there is a superior being to whom the offenders are answerable, and the dreamer has the prerogative and—thanks to texts prescribing such methods—the means to go over the demons' heads and appeal to their overseer. The texts also suggest a semiosis of sorts, a "living sign process through which one thought gives rise to another, which in turn gives rise to another, and so on, into the potential future" (Kohn 2013: 33). That is to say, text and prescribed action alike look to (and help to perform) a near future in which the nightmare-causing agents will have reverted to the control of their superiors, thanks to the semiotic triangulation triggered by the dream and actualized by the wakened dreamer's ritual response. By semiotic triangulation, I mean that the dream is construed as an interaction among three classes of agents. It is by one agent's addressing the other two in the prescribed manner that change is brought about.

To have had a foul dream is to have been thrown, unwittingly and uncontrollably, into a specific sort of interaction with certain unruly other beings. The response is to carry out another, counter-interaction that is deliberate, prescribed, end-directed, and controlled—that is to say, ritualized. The borders of the non-buffered self can thus be restored and fortified.

III. THE SOUL-JOURNEY MODEL

Early on, a model was posited in which dreams occur when one of our multiple souls—cloudsouls (*hun* 魂), whitesouls (*po* 魄), or simply the dreamer's "spirit" (*shen* 神)—wanders outside the body during sleep. The view that we have multiple souls is attested at least as early as the *Zuo Tradition* (a text largely formed by the end of the fourth century BCE). So far as I am aware, the earliest extant passage which unequivocally characterizes soul-wandering as something that happens during sleep, thus strongly implying that it is what accounts for dreaming, occurs in the "Qiwu lun" (齊物論) chapter of *Zhuangzi* (莊子) (ca. 320 BCE?).[6] That the idea is mentioned without elaboration suggests it already had some currency:

> While it [the heart-mind (心)] sleeps, the cloudsouls contact [things] 其寐也魂交
> When it wakes, the bodily form opens up [to sensory input] 其覺也形開
> Whatever we come in contact with entangles it 與接為搆
> Each day we use that heart-mind of ours for strife 日以心鬭.[7]

Similarly, in the "Essence and Spirit" (*Jingshen* 精神) chapter of *Huainanzi* 《淮南子》 (139 BCE), we find it said of the perfected persons (*zhenren* 真人) "whose natures are merged with the Dao" (*xing heyu dao* 性合于道):

> They take life and death to be a single transformation 以死生為一化
> And the myriad things to be a single whole 以萬物為一方
> They merge their essence with the root of great purity 同精於太清之本
> And roam freely beyond the boundless 而游於忽區之旁
> They have vital essence but do not [recklessly] expend it 有精而不使
> They have spirit but do not [carelessly] use it 有神而不行
> They match the artlessness of the great unhewn 契大渾之樸
> And take their stand amid the supremely pure 而立至清之中

Thus their sleep is dreamless 是故其寢不夢
Their wisdom is traceless 其智不萌
Their whitesouls do not sink 其魄不抑
Their cloudsouls do not ascend 其魂不騰[8]

This ancient model of dreaming persisted for centuries, taken up not only in narratives but also in texts instructing readers on how to prevent or limit dreaming by ritually/meditatively anchoring the souls within the body during sleep.[9]

This notion of what dreaming is, widely attested in other societies, presumes a self that is dividual, multipartite, and fissiparous.[10] Dreaming is not (or not just) a mental event: it involves movement, travel, flight by some of our more pneumatic components, parts of us not under our firm control during sleep. In this respect sleep is similar to death, which was also often imagined as the escape of souls from the body and, for some of the souls, a long journey outward, upward, or downward in the cosmos. These were metaphorical models, but they had all sorts of real implications in thought, practice, and narrative.

But even passages as brief as these make clear that a soul's sojourn is not necessarily enough to generate a dream. Rather, dreams arise from encounters or contacts with (*Zhuangzi* writes *jiao* 交 and *jie* 接, the Shangqing Daoist *Purple Texts* speaks of *jiaotong* 交通) other beings during soul journeys. Dreaming itself is a volatile soul's temporary escape during sleep but it is also already an inter-being interaction even before the dreamer wakes to tells others of the dream. Only this time it is a wispy bit of the dreamer's composite self—a mini-me that can wander about meeting other beings on its own steam—that initiates the encounters, to good or (often) ill effect from the perspective of the macro-self being addressed in such texts. Sometimes this macro-self is being chided in such texts for not being a very good manager of its micro-selves, and is offered advice on how to exercise better control.

IV. THE PROSPECTIVE MODEL

A prevalent model of dreams was what I call the prospective model. Here, the point of dreams was that they gave indication of the shape of future events. Dreams and the responses to them, like most activities of sign-making, sign-reading beings, were forward-looking (Kohn 2013: 23–24, 37, 41–42; Seligman et al. 2016). But what they indicated about the future had to be interpreted by the dreamer or someone else, because dreams were thought to be a kind of code, an esoteric language of signs. Someone had to be able to read the signs to yield the correct prediction they indirectly conveyed. On this model, dream interpretation was a kind of divination. By the Han period, the word most often used for those selected aspects of dreams that bore divinable significance (for it was only certain elements of dreams that were factored into interpretations) was *xiang* (象), "simulacra" or "correlates."

There is nothing better than a narrative to afford an example of how such interpretations worked in practice. Whether the events related in the account below actually happened as reported isn't the point; doubtless this record survived thanks to sponsors invested in bolstering Wang Ya's reputation or simply because it was a particularly clever decoding of a dream. My point is that people in the fourth and later centuries often operated with a view of dreams—a dreamscape, as I like to call it, an imaginaire of dreams—that made the following sort of oneiric interpretation imaginable.

The story was related in a now-lost *Annals of Jin* (*Jin Yangqiu* 晉陽秋) by Sun Sheng 孫盛 (fourth century) that was cited in a commentary to *Traditional Tales and Recent Accounts* (*Shishuo xinyu* 世說新語) and related as well in the *History of Jin* and in an early fifth-century collection of anomaly accounts, *A Garden of Marvels* (*Yi yuan* 異苑):

> Zhang Mao (張茂) once dreamed he received an elephant. He asked Wang Ya (王雅) about it. Wang said, "You're about to become governor of a large commandery, but it won't be a good thing. The elephant is a large animal (*da shou* 大獸), which takes the same final sound as 'governor' (*tai shou* 太守). This is how I know you'll be governor of a large commandery. But the elephant loses its life on account of its tusks." Later Zhang did become governor of Wu commandery and was indeed killed by Shen Chong.[11]

The elephant (*xiang* 象) figures twice as a correlate (also *xiang* 象) in Wang's interpretation: for its size, indicating a large administrative unit, and for being killed for its tusks, indicating that Zhang himself will be killed. The way Wang derives Zhang's future governorship is that the elephant is a member of a class (**syuw* 獸 beast) the word for which rhymes with that of the word for the office (**syuw* 守, literally "protector" but part of the compound *taishou* 太守, "grand protector," governor). The story ends, as most such stories do, by noting that subsequent events confirmed the prediction.

Above all, in this and the hundreds of similar instances of dream interpretations under the prospective paradigm, the dream is assumed to be a series of signs, perhaps a mixture of indices and symbols (in Charles Peirce's three-part typology of signs [Kohn 2013: 31–32]). But there is one thing that will seem to many modern readers to be glaringly absent from this account. At no point does Wang Ya ask Zhang Mao about his past, his emotional state, anxieties, desires, or past traumas. This may surprise us, given the dominant modern models of what dreams are, but it wouldn't have surprised medieval readers. For them, a dreamer's emotions, worries, and current preoccupations were recognized as having the potential to shape dreams' contents, but I know of almost no recorded interpretations that focus on such matters. And when these aspects of the dreamer's life situation and psyche are mentioned, it is usually so that they can be factored out of the interpretation—because they were seen as irrelevant at best, distortive at worst. Thus, we find Wang Fu (王符) (90-165 CE) writing: "When one thinks about something during the day and then dreams about it at night, one moment it may appear auspicious and the next moment it may seem inauspicious. The good or bad presage of such a dream is not to be relied upon. This is called a dream that recalls one's thoughts" [畫有所思, 夜夢其事, 乍吉乍凶, 善惡不信者, 謂之想] (*Qianfu lun jian jiaozheng*: 317). Or again, "Only when [the dreamer's] energies have really been touched by something, really informed by spirits, will [accurate] divination be possible" [唯其時有精誠之所感薄, 神靈之所告者, 乃有占爾] (*Qianfu lun jian jiaozheng*: 320, modifying Kinney 1990: 122). Personal—indeed, "psychological"—factors were treated as mere signal interference. The stimulus of a usefully interpretable, prognosticatory dream was supposed to lie outside the dreamer, reflecting not the dreamer's state of mind but the otherwise murky shape of events and the hidden intentions of other actors. For this reason, recorded interpretations rarely reference the psychological state of the dreamer or other matters personal to the dreamer. Instead, interpretations consisted almost entirely of predictions of future events (which might not directly concern the dreamer at all), and were arrived at exclusively through analysis of selected aspects of the dream itself. We could say that dream interpretation was not a psychological process, but instead a cosmologico-semiotic one. It implied a self that was not a container of subjectivity with a determinative past but a multipartite being open to cosmic influences in the present, influences which might leave behind subtle clues to the shape of the future. Those clues could be read by persons skilled in the semiotics of dreams. Dreams were simply one among many kinds of phenomena that afforded clues to the shape of things to come. It was that shape of things that the interpretations revealed and delivered. (Again, the contrast with Freud couldn't be starker: "And what of the value of dreams for our knowledge of the future? Of course that is out of the question. Instead, one should rather ask: for our knowledge of the past. For in every sense, dreams come from the past" [Freud 1999: 412].)

As a semiotic art, dream interpretation was set up analogously to crack divination (*bu* 卜). The dreamer was not thought of as the maker or fashioner of the dream, merely its vehicle. What the dreamer was analogous to was the tortoise carapace or ox scapula. The dream itself was analogous to the crack in the shell or bone. The interpreter of the dream was like the interpreter of the crack. But here the analogy trails off. In the case of dreams, what is it that's analogous to the heated rod that causes the crack, or to the person who applies the rod? If the simulacra comprised in dreams are the counterparts to future events, who or what established those semiotic correlations, and who or what caused the dreamer to dream those *xiang* and not others? Almost all texts concerning dream interpretation are silent on these questions, but that didn't stop dreams from being read.

V. THE VISITATION MODEL

In the fourth chapter of *Zhuangzi* there occurs one of several passages advancing the theme of the usefulness of being useless. A carpenter and his apprentice pass by a shrine shaded by a magnificent old oak tree. The apprentice asks why the carpenter didn't stop to admire the tree, and the carpenter irritably explains that it would be worthless as timber. That night the tree appears to the carpenter in a dream to mount a retort. It holds that perfecting the art of being useless has been very useful to it. The tree continues, "You and I are both things (*wu* 物). What nonsense, that one of us should think it is the other which is the thing! And the worthless human who is about to die, what does he know of the worthless tree?" When the carpenter wakes, he tells of his dream. The apprentice wonders why the tree is standing at the shrine, providing shade and attracting admirers, if it's so intent on being of no use. The carpenter replies:

> Hush! Don't say it. It's simply using that as a pretext, thinks of itself as pestered by people who don't appreciate it. Aren't the ones which don't become sacred trees in danger of being clipped? Besides, it protects itself in a different way from the ordinary. If you try to judge it by conventional standards, you'll be way off![12]

This little story, like many others in *Zhuangzi*, is obviously meant as a parable, not a report of an actual event. But it shares with other narratives built on what I call the visitation model of dreams—most of which were written to be taken as reports of actual events—four important characteristics. First, the dream communicates a message to the dreamer, and it does so directly, not in code—that is, not in *xiang*. Second, the sender of the message is a being who normally can't or doesn't speak to people in waking life. There usually exists a communication gap between trees and humans, but dreaming bridges it, if only temporarily. Third, the message conveyed in the dream brings about a significant alteration in its recipient. Before the dream, the carpenter evaluates the tree as he would any other potential source of material. Afterward he has a new outlook, not only on the tree but, it would seem, on conventional standards generally. The dream encounter changes him. It leaves a real trace in the world. But, fourth, the dream encounter is also itself real; or, to say it differently, dreams are actual interactions with actual other beings.

Consider now the following anecdote, which is attributed to a compilation made around 435. This is an account of events that were alleged to have actually occurred, unlike the *Zhuangzi* story of the oak.[13] A man named Dong Zhaozhi was crossing a river by boat when he noticed a twig floating on the current. An ant scurried back and forth on it, seemingly afraid. So Dong brought the ant on board. That night he dreamed a black-robed man thanked him, saying, "I am king among ants. Should you ever find yourself in trouble, please let me know." Years later Dong was falsely arrested. Remembering his dream, he gathered several ants into his palm and told them of his plight. He then dreamed again of the black-robed personage, who advised him to flee into the hills and await the official pardon that was coming by courier. Waking, he found that ants had chewed through his restraints, allowing him to escape. The pardon soon arrived.

I want to argue that this is not so much a story about dreams as a story about a relationship in which dreams figure importantly. The relationship is built on the exchange of help in dire circumstances. But it begins in an inter-subjective encounter in which the human protagonist recognizes another self, a self that recognizes itself to be in trouble on the water. Significantly, some of the textual variants that have come down to us say that as it scurried to and fro on the twig the ant "feared for its life" [惶遽畏死]; the ant thus becomes a narrative subject, a co-protagonist.[14] Other variants put this as an inference in the mind and speech of Dong: "Zhaozhi said [to himself], 'This means it is in fear for its life' " [昭之曰此畏死也]; the ant is thus apprehended by the man, via what philosophers are wont to call a "theory of other minds," as a living subject like himself, whose behavior evinces an awareness of danger and the aim to stay alive. It is a story, then, of the emergent relationship between two selves—a narrative of an ecology of selves (Kohn 2007; 2013). Each of these protagonists has aims—most especially the aim to keep on living and thriving. Each recognizes the other as an intentional being with aims similar to its own, and acts accordingly.

The two selves fall into a form, a pattern of gifting, which, once entered into, interlinks both parties even if it does not tightly bind them.[15] It is a form, the story invites us to think, that transcends species. It's not a conceptual or merely cultural structure imposed by the human party on the blank canvas of "nature" so much as an emergent feature of the process of interaction between the two selves, constraining both of them. We might even read the story as inviting us to consider whether, rather than a nature/culture binary in which cultures are plural but nature is singular, there is in fact only one culture—in which forms such as that of gift-giving link selves of diverse species—but many natures, in the sense of many *Umwelten*, many worlds-as-experienced depending on each species' distinctive sensorimotor capacities and how they shape its perception of and relation to their environment.[16]

Some would read this story as a charming projection of a uniquely human, cultural, and perhaps "Confucian" value (that of *bao* 報 or moral reciprocity), and of the uniquely human process of sign-making onto the sign-less, value-less, self-less, aim-less nonhuman world of nature. In such a reading, the story would have only human beings and human culture to be about. Some historians of Chinese literature might simply read the story as an instance of the "birth of fiction," in which case it has only the prehistory of a literary genre to be about. But I want to pursue a different reading. As Eduardo Kohn has argued,

> The distinction . . . is not between an objective world, devoid of intrinsic signifi-cance, and humans who, as bearers of culture, are in a unique position to give meaning to it . . . . Rather, "aboutness"—representation, intention, and purpose in their most basic forms—emerges wherever there is life; the biological world is constituted by the ways in which myriad beings—human and nonhuman—perceive and represent their surroundings. Significance . . . is not the exclusive province of humans (Kohn 2007: 5).[17]

And the narrative's two dream events are crucial to this reading.

Dreaming is a privileged mode of cross-kind communication, allowing the ant—normally a user of indexical signs (to use Peirce's tripartite semiotic typology[18])—to use symbolic signs, that is, human language, as well.[19] That, we could say, is one affordance of dreaming.[20] Another is the opportunity for the ant to appear in human form and garb. Selves come clothed in bodies, and in dreams the clothing may be changed. Dong presumably would not have understood ant signs, but the king of ants is able to cross the threshold and meet him more than halfway in dreams. This oneiric affordance usually moves in only one direction (cf. Kohn 2013: 167): Dong does not appear in the ant's dream, addressing him in ant signs and wearing the physical form of an ant. Maybe that's simply because the record we have to read of the event is a record made by humans.

Dreaming, then, among other things afforded a portal for face-to-face encounters with others across ontological, taxonomic, spatial, and linguistic gaps. It afforded relationships. Note that it's not as if the human protagonist at once recognized the ant as a self with aims because that's how Chinese people viewed the class of beings known as ants. In many situations, an ant might have gone unnoticed entirely. In others, it might have been seen as an object, an instrument, or a nuisance. (In fact, some versions mention that other passengers on Dong's boat complained about his bringing a biting creature aboard and threaten to crush it. Out of pity he carefully shields the ant until the boat is docked.) In this instance, its being seen as a self with aims emerges from the interaction of these two beings in their specific situation on the river. The ant scurries back and forth on the floating stick; the human being notices this unusual behavior, sizes up the situation, infers the ant's aims, and extends help. It's this pairing of human attention and action with the purposive behavior of the distressed ant that leads into a relationship that, facilitated by the communicative portal of dreaming, stretches over years (cf. Bird-David 1999: 75).

Rather than being private and isolative, then, dreams here again connected the dreamer to other selves, and even more intimately. Dreaming was conceived of as an interpersonal space—and the persons who might find each other there were not limited to living human persons. This is a view of dreaming that may surprise us, but it turns out to be quite common, widely distributed in human history and geographic space. Kevin Groark captures it well:

> In many traditional societies, the dreamspace forms an alternate interpersonal sphere characterized by forms of social interaction and experience that are qualitatively different from waking life, yet intimately related to it . . . . Among the Tzotzil Maya, human sociality is explicitly understood to consist of both face-to-face relations (relations between physical selves) as well as relations between various self-extensions or soul-based "counterparts." The dreamspace serves as an interpersonal realm in which the normally occluded motives and feelings of others (both human and extra-human) can be perceived and experienced through the medium of soul interactions . . . . Dream experience allows for a translocation or shift in the focus of attention and intersubjective engagement from the relatively opaque phenomenal realm of physical bodies to the essential realm of souls. In doing so, it brings the extended intersubjective field of soul-based "counterpart relations" . . . into focus as part of the total field in which relationality is embedded. (Groark 2013: 285, emphasis added)

The limit-case of the visitation model might be stories in which the protagonist actually becomes, through dreaming, an extra-human self. The most famous example is surely the *Zhuangzi*'s parable of the butterfly dream, but the following incident from *Miscellaneous Morsels from South of You* (*Youyang Zazu* 酉陽雜俎), compiled by Duan Chengshi (段成式) (803–863), was strongly claimed to have actually occurred, unlike the *Zhuangzi* tale.

A man named Han Que had long savored pickled minced fish. He sent a servant to bring him some fish from a dam at a local lake. He then fell asleep and "dreamed he himself was a fish, forgetting himself in delight in the lake" [夢身為魚, 在潭有相忘之樂]. Soon he experienced being netted by fishermen, flung into a pail, taken to market, and sold to the very servant whom he had sent to buy fish. The servant lifted him by the gills and strung him through the throat. When the servant arrived home with him, Han Que recognized his own wife, children, and servants. Someone put him on a chopping block and cut him in two. The pain was as if he were being skinned alive. Only when his head fell did he awaken. "His consciousness was muddled for a long time" [神凝良久]. When family members questioned him, he recounted the dream to them. He then summoned the servant and quizzed him on the details of his trip to buy the fish. "What [the servant] recounted was what he had dreamed" [具述所夢]. Han Que responded by joining the Buddhist sangha. To this entry, Duan Chengshi adds a note that affords us a glimpse of the interpersonal networks along which such anecdotes traveled before being set down in the written compilations where we now typically find them: "The home of my scribe, Chen Zhi, is in Yuezhou, close to the dam, and he himself witnessed this affair."[21]

Here again, the oneiric experience leaves a deep trace in the waking world, spurring the dreamer to renounce household life and take monastic vows. "Entertaining the viewpoints of other beings is dangerous business" indeed (Kohn 2007: 7)—dangerous but also instructive and potentially fruitful.

VI. CONCLUSION

Dreaming is a common experience. Yet it's also an intrinsically strange one.

> What's most important about dreaming is that it allows you to experience a world where the normal waking rules don't apply, where causality and rational thought and our core cognitive schemas (people don't transform or merge, places should be constant, gravity always operates, and so forth) melt away in the face of bizarre and illogical stories. And, while you dream, you accept these stories as they unfold. Essentially, the experience of narrative dreams[22] allows you to imagine explanations and structures that exist outside of your waking perception of the natural world. In your waking life you may embrace the distorted structures of the dream world or you may be a hard-headed rationalist, or you may blend the two (as most of us do), but in all cases the experience of dreaming has thrown back the curtain and allowed you to imagine a world where fundamentally different rules apply. (Linden 2007: 220)

In societies where lavish energy is spent training bodies and minds to produce humans with the correct habits and comportment dreaming can serve as an escape hatch. It can be what Jane Bennett calls a crossing, opening up new possibilities and bringing new things into being.[23] Bennett writes:

> Norbert Elias has chronicled . . . practices as they were inculcated through books of manners from the fifteenth through the eighteenth centuries. He shows how

human animals engaged in painstaking attempts to regulate their hand, eye, and mouth movements while eating, drinking, sharing a bed, urinating, spitting, defecating, and conversing . . . . The overall effects of these exercises were (1) to confine bodily movements to those that mimicked the delicacy of the courtly class and (2) to extend the psychological and phenomenological distance between human and animal. (2001: 25)

China had its versions of these sorts of exercises, notably the behaviors, norms, and traditions Confucians summarily called "ceremony" (*li* 禮). Dreaming remained one area of experience resistant to this regimen of habituation (despite evidence that some Confucians tried to domesticate even dreams for self-cultivation purposes[24]), which is why the authors of texts such as *Zhuangzi* and *Liezi* (列子) often turned to dreams to remind readers of the limitedness of any point of view and to leverage the possibility of transformation, of wandering beyond the carefully ruled grid of enculturation: in short, the possibility of enchantment. They found dreams useful to write about and good to wield in arguments about reality and knowledge because, as anthropologist Waud Kracke has observed, "dreams are the most obstinate stumbling block to our secure knowledge that we 'know' the world around us, that we all share the same experience of it and of each other" (2003a: 212).

By attending more deliberately to our dreams in the midst of the Anthropocene, we might, as did the character Zhuangzi when he woke from a dream in which he'd been a butterfly, find a bit of wiggle room for enchantment, for realizing "what's called the transformations of things" (此之謂物化) (*Zhuangzi jijie* 27). But we can also try inhabiting different models of dreaming—and thus different models and senses of ourselves—than those modern ones that encourage us to see ourselves as the sole makers, sources, and topics of dreams, and to think of ourselves as radically separate and distinct from the rest of the world we inhabit. As Kracke writes:

Much of the world has been talking about dreams for centuries; it stands to reason that such cultures may have as much to tell us about dreams as pharmacognicists are now recognizing that indigenous cultures can teach us about pharmacology. If Amazonian cultures may open up botanical secrets to the prepared tropical botanist . . . then the multitude of cultures that have taken dreams seriously while we dismissed them as "froth" or "random neuronal firing" may well be able to amplify our vision of dreams. (2003b: 157)

Opening ourselves to other models of dreaming, such as those from late classical and medieval China, might open us to de-anthropocentrizing understandings of ourselves and alternate ways of experiencing ourselves as related to other things. It might move us toward inhabiting and practicing a new animism—that is to say, a relational epistemology (Bird-David 1999: 68, 73), or "the attribution by humans to nonhumans of an interiority identical to one's own" (Descola 2013: 129), or the recognition "that the world is full of persons, only some of whom are human, and life is always lived in relationship with others" (Harvey 2017: 9).

NOTES

1. *Zhuangzi jijie*, 41–42, partially adopting the translations in Watson (1968: 63–65) and Graham (1981: 72–73).

2. An important start in this direction was made in Hall (1996), and it has been followed by others, but most studies of this sort of which I'm aware are small-scale, survey-based statistical reports. We await a large-scale interpretive work with a broad historical perspective.

3. For the template's "so-and-so" (*mou* 某), the dreamer is to substitute his or her own name.

4. There is a pun here between the text's "broadcloth" (*fu* 幅) and "blessings" (*fu* 福) and "wealth" (*fu* 富); quality textiles were themselves, of course, a form of wealth.

5. Translation is that of Harper (2010: 52); cf. Harper (1988: 72–73). Photographs, transcriptions, and annotations of the Shuihudi daybook passages may be consulted in Yu Haoliang et al. (1990), plates 134–35 and pages 210, 247.

6. In view of Klein (2011), this dating may be too early, in which case we would not, to my knowledge, have firmly datable textual evidence preceding the Western Han that dreaming was seen as spirit travel.

7. *Zhuangzi jijie* 11; translation adapted from Graham (1981: 50).

8. *Huainanzi jishi* 524–25, modifying Major et al. (2010: 249).

9. Such as in the Shangqing *Purple Texts*, translated in Bokenkamp (1997: 275–372).

10. The description given by Poirier (2003: 114) of Australian aboriginal views of the self as it relates to dreaming applies equally here: it is a "notion of the person as separable . . . as permeable to and consubstantial with non-human agents and essences, and composed of multiple relationships (that some might call 'identities') . . . . These relationships . . . are intrinsic to the bodily self . . . . In other words, they are constitutive of one's personhood . . . . The local notion of the person is . . . 'dividual.' "

11. Translation modified from Mather (2002: 271). See also Campany (2015: 94–95).

12. *Zhuangzi jijie*, 41–42, partially adopting the translations in Watson (1968: 63–65) and Graham (1981: 72–73).

13. The work in question is *Xie Qi's Records* (*Qi Xie ji* 齊諧記), attributed to Dongyang Wuyi (東陽无疑) (fl. ca. 435). This story is translated in Campany (2015: 24–25), and the textual sources for it (variants among which are discussed here) are listed in ibid., 136.

14. When mentioning variant versions of the same anecdotes, I do so on the assumption that each version that has come down to us was an artifact of social memory; each therefore has value in assessing what people claimed to have happened, or what they thought might possibly happen. It's not a question, then, of trying to ascertain which version was "original" or "correct." For more on this point, see Campany (2012: 17–30) and works cited there.

15. Following Descola (2013: 307–335), I see the relationship here as one involving gifting, not exchange. "Unlike exchange, the gift is above all a one-way gesture that consists in abandoning something to someone without expecting any compensation other than that, possibly, of gratitude on the part of the receiver . . . . Reciprocal benefaction is never guaranteed where a gift is concerned" (313). But I would add that in China, the expectation of a commensurate return has often been stronger than Descola suggests. On the one hand, then, Dong's initial gesture was not done on the expectation of a return; but on the other hand, the ant king's response to it was, in Chinese context, hardly unexpected, either (except in the obvious way that he was a particular ant with extraordinary power to be of assistance).

16. "Because an organism is the locus of the work that is responsible for generating the constraints that constitute information about its world, what this information can be about is highly limited, specific, and self-centered. Like the treasure hunter with his metal detector, an organism can only obtain information about its environment that its internally generated dynamic processes are sensitive to—von Uexküll's *Umwelt*, the constellation of self-centered species-relevant features of the world" (Deacon 2013: 410). See also Hoffmeyer (2008: 171–211); Wheeler (2006: 120–122); Uexküll (2010).

17. Compare: "Processes of sign and meaning cannot, as is often assumed, become criteria for distinguishing between the domains of nature and culture. Rather, cultural sign processes must be regarded as special instances of a more general and extensive biosemiosis that continuously unfolds and acts in the biosphere" (Hoffmeyer 2008: 4).

18. A sign, for Peirce, is "something which stands to somebody for something in some respect or capacity" (Kohn 2013: 29; Hoffmeyer 2008: 20). Peirce distinguished three general classes of signs. Icons are usually likenesses of their objects (e.g., photographs). Indexes, rather than being likenesses of the objects they represent, point to them (e.g. weathervanes as indices of wind direction). Symbols refer to their objects indirectly as a function of their systematic and conventional relationship to other symbols (e.g. words in a human language). "Unlike iconic and indexical modes of reference, which form the bases for all representation in the living world, symbolic reference is, on this planet at least, a form of representation that is unique to humans" (Kohn 2013: 31–32).

19. Compare: "Dreaming is understood to be a privileged mode of communication through which, via souls, contact among beings inhabiting different ontological realms [that is, different Umwelten] becomes possible" (Kohn 2007: 12).

20. On the notion of affordances as intended here and below, see Gibson (1977); Levine (2015: 6–11); Bird-David (1999).

21. My summary draws on Reed (2003: 136). The story is internally dated to 837.

22. By this Linden means the long, narratively complex sort of dream typical of REM sleep shortly before waking, as opposed to (1) brief, sensorily rich but non-narrative dreams typical in the period just after sleep onset and (2) dreams typical of deeper, non-REM sleep, which may be heavily laden emotionally but are nonetheless non-narrative (Linden 2007: 209–211).

23. Bennett uses "crossings" as a collective noun roughly synonymous with "hybrids" (2001: 17–32). I intend it to mean a type of transformative process that, like hybrids, "in some cases retains the power to open a window onto previously occluded lines of alliance, affiliation, and identification" (32).

24. I have in mind specifically *Analects* 7.5 and other early passages that portray Confucius as receiving answers to questions in his dreams or else lamenting that he no longer does so. See, for example, Knoblock and Riegel (2000: 618).

ENGLISH AND WESTERN LANGUAGES

Bennett, Jane. 2001. *The Enchantment of Modern Life: Attachments, Crossings, and Ethics*. Princeton: Princeton University Press.

Bird-David, Nurit. 1999. "'Animism' Revisited: Personhood, Environment, and Relational Epistemology." *Current Anthropology*, vol. 40, special issue 1, pp. 67–91.

Bokenkamp, Stephen R. 1997. *Early Daoist Scriptures*. Berkeley: University of California Press.

Campany, Robert F. 2003. "On the Very Idea of Religions (in the Modern West and in Early Medieval China)." *History of Religions*, vol. 42, pp. 287–319.

Campany, Robert F. 2012. *Signs from the Unseen Realm: Buddhist Miracle Tales from Early Medieval China*. Honolulu: University of Hawai'i Press.

Campany, Robert F. 2015. *A Garden of Marvels: Tales of Wonder from Early Medieval China*. Honolulu: University of Hawai'i Press.

Deacon, Terrence W. 2013. *Incomplete Nature: How Mind Emerged from Matter*. New York: Norton.

Descola, Philippe. 2013. *Beyond Nature and Culture*. Translated by Janet Lloyd. Chicago: University of Chicago Press.

Freud, Sigmund. 1999. *The Interpretation of Dreams*. Translated by Joyce Crick, with an introduction by Ritchie Robertson. Oxford: Oxford University Press.

Gibson, J. J. 1977. "The Theory of Affordances." In *Perceiving, Acting, and Knowing*, edited by R. E. Shaw and J. Bransford, pp. 67–82. Hillsdale, NJ: Lawrence Erlbaum Associates.

Graham, A. C. 1981. *Chuang-tzu: The Inner Chapters*. London: George Allen & Unwin.

Groark, Kevin. 2013. "Toward a Cultural Phenomenology of Intersubjectivity: The Extended Relational Field of the Tzotzil Maya of Highland Chiapas, Mexico." *Language and Communication*, vol. 33, pp. 278–291.

Hall, David H. 1996. "Beliefs about Dreams and their Relationship to Gender and Personality." PhD dissertation, Wright Institute Graduate School of Psychology.

Harper, Donald. 1988. "A Note on Nightmare Magic in Ancient and Medieval China." *T'ang Studies*, vol. 6, pp. 69–76.

Harper, Donald. 2010. "The Textual Form of Knowledge: Occult Miscellanies in Ancient and Medieval Chinese Manuscripts, Fourth Century B.C. to Tenth Century A.D." In *Looking at It from Asia: The Processes that Shaped the Sources of History of Science*, edited by Florence Bretelle-Establet, pp. 37–80. Paris: Springer.

Harvey, Graham. 2017. *Animism: Respecting the Living World*. 2nd edition. New York: Columbia University Press.

Hoffmeyer, Jesper. 2008. *Biosemiotics: An Examination into the Signs of Life and the Life of Signs*, translated by Jesper Hoffmeyer and Donald Favareau, edited by Donald Favareau. Scranton and London: University of Scranton Press.

Josephson-Storm, Jason Ā. 2017. *The Myth of Disenchantment: Magic, Modernity, and the Birth of the Human Sciences*. Chicago: University of Chicago Press.

Jung, Carl G. 2010. *Dreams*. Translated by R. F. C. Hull. Princeton: Princeton University Press.

Kinney, Anne Behnke. 1990. *The Art of the Han Essay: Wang Fu's Ch'ien-fu lun*. Tempe: Center for Asian Studies, Arizona State University.

Klein, Esther S. 2011. "Were There 'Inner Chapters' in the Warring States? A New Examination of Evidence about the *Zhuangzi*." *T'oung Pao*, vol. 96, pp. 299–369.

Knoblock, John, and Jeffrey Riegel. 2000. *The Annals of Lü Buwei: A Complete Translation and Study*. Stanford: Stanford University Press.

Kohn, Eduardo. 2007. "How Dogs Dream: Amazonian Natures and the Politics of Transspecies Engagement." *American Ethnologist*, vol. 34, no. 1, pp. 3–24.

Kohn, Eduardo. 2013. *How Forests Think: Toward an Anthropology beyond the Human*. Berkeley: University of California Press.

Kracke, Waud. 2003a. "Afterword: Beyond the Mythologies, a Shape of Dreaming." In *Dream Travelers: Sleep Experiences and Culture in the Western Pacific*, edited by Roger Ivar Lohmann, pp. 211–235. New York: Palgrave Macmillan.

Kracke, Waud. 2003b. "Dream: Ghost of a Tiger, a System of Human Words." In *Dreaming and the Self: New Perspectives on Subjectivity, Identity, and Emotion*, edited by Jeannette Marie Mageo, pp. 155–164. Albany: State University of New York Press.

Kunzendorf, Robert G., et al. 2007–2008. "The Archaic Belief in Dream Visitations as it Relates to 'Seeing Ghosts,' 'Meeting the Lord,' as well as 'Encountering Extraterrestrials.'" *Imagination, Cognition and Personality*, vol. 27, no. 1, pp. 71–85.

Latour, Bruno. 1993. *We Have Never Been Modern*. Translated by Catherine Porter. London: Prentice Hall.

Levine, Caroline. 2015. *Forms: Whole, Rhythm, Hierarchy, Network*. Princeton: Princeton University Press.

Linden, David J. 2007. *The Accidental Mind*. Cambridge, MA: Harvard University Press.

Major, John S., Sarah A. Queen, Andrew Seth Meyer, and Harold D. Roth. 2010. *The Huainanzi: A Guide to the Theory and Practice of Government in Early Han China*. New York: Columbia University Press.

Mather, Richard B. 2002. *Shih-shuo Hsin-yü, A New Account of Tales of the World*. 2nd edition. Ann Arbor: Center for Chinese Studies, University of Michigan.

Mazandarani, Amir Ali, Maria E. Aguilar-Vafaie, and G. William Domhoff. 2018. "Iranians' Beliefs about Dreams: Developing and Validating the My Beliefs About Dreams Questionnaire." *Dreaming*, vol. 28, no. 3, pp. 225–234.

Morewedge, Carey K., and Michael I. Norton. 2009. "When Dreaming is Believing: The (Motivated) Interpretation of Dreams." *Journal of Personality and Social Psychology*, vol. 96, pp. 249–264.

Perls, Frederick. 1969. *Gestalt Therapy Verbatim*. Compiled and edited by John O. Stevens. Moab, UT: Real People Press.

Poirier, Sylvie. 2003. "'This is Good Country. We Are Good Dreamers': Dreams and Dreaming in the Australian Western Desert." In *Dream Travelers: Sleep Experiences and Culture in the Western Pacific*, edited by Roger Ivar Lohmann, pp. 107–125. New York: Palgrave Macmillan.

Puett, Michael J. 2012. "Social Order or Social Chaos." In *The Cambridge Companion to Religious Studies*, edited by Robert A. Orsi, pp. 102–129. Cambridge: Cambridge University Press.

Reed, Carrie E. 2003. *A Tang Miscellany: An Introduction to "Youyang zazu."* New York: Peter Lang.

Rock, Andrea. 2004. *The Mind at Night: The New Science of How and Why We Dream*. New York: Basic Books.

Seligman, Martin E. P., and Peter Railton, Roy F. Baumeister, and Chandra Sripada. 2016. *Homo Prospectus*. New York: Oxford University Press.

Slingerland, Edward. 2019. *Mind and Body in Early China: Beyond Orientalism and the Myth of Holism*. New York: Oxford University Press.

Taylor, Charles. 2011. "Western Secularity." In *Rethinking Secularism*, edited by Craig Calhoun, Mark Juergensmeyer, and Jonathan VanAntwerpen, pp. 31–53. New York: Oxford University Press.

Uexküll, Jakob von. 2010. *A Foray into the Worlds of Animals and Humans, with A Theory of Meaning*. Translated by Joseph D. O'Neil. Minneapolis: University of Minnesota Press.

Watson, Burton. 1968. *The Complete Works of Chuang Tzu*. New York: Columbia University Press.

Wheeler, Wendy. 2006. *The Whole Creature: Complexity, Biosemiotics and the Evolution of Culture*. London: Lawrence & Wishart.

CHINESE LANGUAGE

Haoliang, Yu 于豪亮 et al., eds. 1990. *Shuihudi Qin mu zhujian* 睡虎地秦墓竹簡. Beijing: Wenwu chubanshe, 1990.

Huainanzi jishi 淮南子集釋. Edited by He Ning 何寧. 3 vols. continuously paginated. Beijing: Zhonghua shuju, 1998.

Qianfu lun jian jiaozheng 潛夫論箋校正. Edited by Peng Duo 彭鐸, annotated by Wang Jipei 汪繼培. Beijing: Zhonghua shuju, 1985.

Zhuangzi jijie. 莊子集解. Edited by Wang Xianqian 王先謙 and Liu Wu 劉武. Beijing: Zhonghua shuju, 1987.

Chapter 6

A Syncretic Innovation in Chinese Buddhism

Animal Release Rituals in New York City

Wei Dedong[1]

When an independent national culture encounters another culture of a different kind, it has three possible paths forward: 1) engage in self-appreciation while rejecting any engagement with the Other; this results in isolation and sealing itself off from the world; 2) accept assimilation, give up its own original culture, assiduously imitate the Other culture – this results in losing one's cultural independence and becoming the servant of the stronger state; 3) actively learn from and integrate the foreign culture, and grasp its essence, so that one's national culture can be strengthened. When traditional Chinese national culture encounters modern Western culture, it should adopt the attitude of the third path.

—(Zhang Dainian and Cheng
Yishan 2015: 63)

Most elites in late imperial China and especially during the late Qing Dynasty chose the first path described in the epigraph of this chapter. Since the May Fourth Movement (1919–1930s), Chinese reformers and revolutionaries favored the second path. When the Chinese philosopher Zhang Dainian (張岱年) developed his Theory of Syncretic Innovation (綜合創新論) in the 1990s, this third path was a breakthrough that answered the question of how to modernize traditional Chinese philosophy and culture (Zhang Dainian and Cheng Yishan 2015). As shown in the epigraph, Zhang's method was to retrieve the different roots of traditional Chinese philosophies, and integrate all the outstanding achievements of ancient, modern, Asian and Western civilizations, and create a novel and modern form of Chinese philosophy. After a century of zigzagging from one extreme of cultural isolationism and attitude of cultural superiority, to another extreme of "complete Westerniza-tion" (全盤西化) and abandonment of traditional Chinese culture, Zhang's intervention in the 1990s' "Cultural Fever" (文化熱) debates was to advocate cultural hybridity and cosmopolitanism.

Zhang's Theory of Syncretic Innovation can also apply to the modern transformation of traditional Chinese Buddhism. In the twenty-first century, the creative adaptation of the ancient Buddhist ritual practice of "Releasing Life" (*fangsheng* 放生) in New York City is one product of the syncretic innovation of traditional Chinese Buddhism. The traditional practice is based on the Buddhist idea of releasing animals that are to be killed for food and that were either domesticated or captured wildlife, to accumulate spiritual merit. The syncretic innovation adapts rituals, cooperates with animal reha-bilitation and ecological protection in an urban context to transform the tra-ditional approach of releasing animals into a new "wise" release that is based upon scientific knowledge of animals, and also contributes to contemporary environmental protection.

From 2016 to 2019, I conducted fieldwork and engaged in participant observation of Chinese Buddhist "releasing life" rituals, mainly turtles and birds, in the City of New York. I collected firsthand data and conducted in-depth personal interviews with the leaders of the Grace Gratitude Buddhist Temple (GGBT), the New York Turtle and Tortoise Society (NYTTS), the Wild Bird Fund (WBF), Central Park in Manhattan, and the Jamaica Bay Wildlife Refuge in Queens. The leaders of the NYTTS and the WBF were the main advocates of a "wise release of life." When Venerable Benkong stayed at my home in Beijing for a month in 2019, I was able to further interview him at length.

This chapter will first define and explain the concepts of the Buddhist "release life" rituals, and what I call the "blind release of life" (盲目放生) and the "wise release of life" (智慧放生), and then it introduces the Buddhist teachings on "releasing life." Next, the chapter analyzes the "blind release of life" in New York City, and then considers the recent developments of the "wise release of life" by Chinese Buddhists residing in New York City, where many Buddhists are first generation immigrants. Finally, the chapter will discuss the significance of the "wise release of life," not only for Chinese Buddhism, but also for modern societies around the globe.

I. RELEASE OF LIFE, BLIND RELEASE OF LIFE, AND WISE RELEASE OF LIFE

"Releasing life" (*fangsheng* 放生) is one of the most popular practices of modern Buddhism that has started to be disseminated from first, Taiwan in the

1970s, then from the Sinophone world in Southeast Asian societies, then from mainland China in the 2000s, to all over the world. The purpose of the ritual of "releasing life" is to rescue animals who are in captivity and in distress as they wait to be slaughtered, and return them to the wild, while ensuring their safety. In Buddhist rituals of releasing life, the human Buddhist participants also educate the animals about Buddha doctrines, which prepares them to "seek the three refuges" (三皈依) (in the Buddha, the Dharma, and the Sangha), so that someday in the future, the animals may also become Buddhist disciples. After "taking refuge," the animals will have the potentiality to become a buddha and be able to themselves express the great compassion of Buddhism. Many Chinese Buddhists also believe that participation in "releasing life" rituals may bestow on them a cure for their illness and ensure their longevity.

The fundamental driving force behind a Buddhist's participation in the ritual of "release of life" is the accumulation of Buddhist merit. However, if people only pay attention to the purpose of accumulating personal merit for themselves in this or their next lives, without caring for the animals and the environment, they may be participating in a kind of "Blind Releasing Life" ritual. This blind release may actually lead directly to animal deaths, destroy ecological balance and harmony, and corrupt the reputation of Buddhism in modern society, so it loses the function of accumulating merit.

In China, as early as the Ming Dynasty, the Buddhist monk Master Lianchi (蓮池 1535-1615 CE) already criticized the phenomenon of "blind animal release". According to Master Lianchi in *Ahimsa and Releasing Life Texts*《戒杀放生文》:

> People may buy animals, keep them for one night, and release the next morning; or they buy in the morning and keep them for the afternoon. If so, organizers must furnish the ritual site and assemble many men and women, so the time is too long and half the animals will die. This kind of release of life is just an empty name.
> 若隔宿買而來朝始放，或清晨買而午后猶存，必待陳設道場，會集男女，迁遷延時久，半致死亡。如是放生，虛文而已.

Lian Chi（2012: 1288）

The "blind release of life" also grew into a profitable industry in China, starting in late imperial times, and in its recent revival in the highly commercialized society of contemporary China, it has become even more of a problem. Since the time, place, and animal species of Buddhist release rituals are often fixed and generally known to the public, some unscrupulous profit-minded people go out into the wild to capture animals such as birds, fish, and turtles, specifically to sell to devout Buddhists for this ritual. In the process of capturing animals from the wild, transporting them in densely packed crates or tubs to the market place, and releasing them back into the wild, innumerable animals and fish perish. They die in the harsh conditions of transport, or they perish when they are carelessly reintroduced into an alien environment unsuited to their survival. Many are the times when careless or uninformed Buddhists release freshwater fish into the ocean, mountain birds from one part of the country onto the dry plains of another region, or tropical turtles into northern lakes that freeze over in the winter. Furthermore, the natural environment may also be seriously damaged as a result of so many eager ritual participants crowding into an ecologically fragile environment. After a century in which Buddhist rituals were often attacked as "superstitious" and "backward," many new Chinese Buddhists today seem to focus more on their personal salvation than on the welfare of the animals they are releasing. Furthermore, their lives in densely populated urban surroundings meant that many people have lost touch with the basics of nonhuman species survival in the wild.

With the introduction of Chinese Buddhism into the United States, the "blind releasing of life" has also entered, including its form of ritual that has now often lost the intended Buddhist compassion for sentient beings. Beginning around 2010, Chinese-American Buddhists in New York City's Chinatown developed a new form of the "release life" ritual in cooperation with local animal rehabilitation organizations and ecological protection zones, achieving an ideal combination of traditional Buddhist practices of merit accumulation, modern scientific animal rehabilitation, and ecological balance. This is what I call the "Wise Release of Life".

Three persons in the New York City area have played key roles in this process of inventing the "wise release of life": Master Jingyi (釋淨義) a Chinese Buddhist nun and the abbess of Grace Gratitude Buddhist Temple (GGBT) in Manhattan's Chinatown; Venerable Benkong (釋本空), a Euro-American Buddhist senior monk at GGBT; and Lorri Cramer, the Director of the Turtle Rehabilitation and Curriculum Development in the NYTTS.

Master Jingyi was born in Fujian Province, China, and became a nun when she was sixteen years old in 1978. She used to be the leader of the "release life" team at the Fujian Provincial Buddhist Association. She came to the United States in 1995 and became the abbess of the GGBT in 2004. Based on her experience of releasing animals in China, Master Jingyi continued to be enthusiastic about the ritualistic releasing of animals after she arrived in the United States, until one year, when she received a stern warning by the Boston District Court that her actions violated City regulations regarding the protection of animal welfare. She began thinking about how to introduce innovations into the traditional releasing methods, to ensure that the released animals would survive their reintroduction into the wild. She and her temple finally embarked on the "Wise Release of Life" path in cooperation with

New York City animal rehabilitation institutions and ecological protection zones.

Master Jingyi's passion for releasing animals has something to do with her personal religious experience. She told me a story about a "release life" ritual in her hometown in Fujian Province that occurred in 1993:

> One day, I spotted someone selling a pangolin on the street. I went out to raise money to buy the pangolin and set it free on the mountain behind my temple. The next morning, I noticed a pile of earth behind the temple as I was building a new house next to it. The earth was just what I needed for the day's construction work. While I was working on constructing the house for two weeks, the fresh new pile of earth appeared every day. Finally, I said, "Pangolin, thank you for helping me to release the earth. I have now finished building my house, so I don't need the earth anymore. You don't have to work so hard anymore." The very next day, there was no more earth pile behind the temple.

For these reasons, Master Jingyi firmly believes that animals have the ability to understand and communicate with human consciousness. For Master Jingyi, this means that all living beings possess "Buddha Nature" (佛性), and have the ability to become buddhas one day.

Venerable Benkong (originally known as Harold Lemke) is a native of Jersey City, New Jersey, United States. When he was a child, his grandmother lived in Little Italy, which was located right next to Chinatown in New York City. Thus, since childhood, Venerable Benkong often went to Chinatown to play, thereby forming a fate to live a life deeply involved with Chinese culture. In 1969, seventeen-year-old Harold Lemke journeyed to Taiwan to learn Chinese language and culture. He studied at an American high school at first, and then entered into the Department of History at Fu Jen Catholic University in Taiwan. He also studied at Seton Hall University and other colleges. After his student life, he became a translator in Taiwan and San Francisco for about ten years. Then he went to South Africa to set up an AIDS prevention nongovernmental organization, called the Centre for Positive Care, and worked for ten years there. In South Africa, he encountered Chinese Buddhism, became deeply immersed in the study of Buddhist sutras and teachings, and he was ordained a Buddhist monk at the age of forty-nine, with the Buddhist name of Shi Benkong. The greatest advantage that he enjoyed is that he is proficient in English, Chinese, and Buddhist Dharma, and therefore, he is well-positioned to make a special contribution to the spread of Buddhism to the American people.

In the process of the innovation of the "wise release of life," Venerable Benkong mainly shoulders the work of communicating with the animal protection organizations, the environmental institutions, and the lay Buddhists. According to Iris Ho of the Humane Society International, "Over the past

couple of years, Venerable Benkong has been working with and supporting licensed turtle rehabilitators' work in rescuing, nurturing and caring for
injured turtles and tortoises, and releasing healthy ones back into the wild"
(Ho 2012). Venerable Benkong said, "My ultimate goal is for every Buddhist
temple in the United States to have a rehabber or conservation group that they
support and use to educate their community. They just need to come knock
on our door." Thus, Venerable Benkong offered the interpretation of wise
releasing both in line with the spirit of Buddhism and the social environment
of the United States.

Another key person who helped develop the "wise release of life" is Ms.
Lorri Cramer, NYTTS Director of Turtle Rehabilitation and Curriculum
Development, and affectionately called the "Turtle Lady." She developed
her expertise rehabilitating turtles over some thirty-four years, nurturing over
1,000 turtles and tortoises back to health, and releasing them back into the
wild. In her work of rescuing injured or ill turtles in the New York City area,
Ms. Cramer discovered that someone or some group was repeatedly releasing turtles into inappropriate places, jeopardizing their survival. The releases
were traced to Chinese Buddhist Life Release ceremonies. She wrote many
letters to Chinese Buddhist temples, and eventually, "Ms. Cramer was able
to make successful contact with an important member of the New York Chinese Buddhist community" (Lombardi 2014). One day, she received a call
from Venerable Benkong of GGBT, and thus, the cooperation between the
animal rehabilitation organizations and the Buddhist groups began (figures
6.1 and 6.2).

Only in the past ten years, beginning around 2010, did the ancient Buddhist ritual practice assume the new form of the "wise release of life." This
Buddhist ritual now includes four steps: (1) the GGBT or other Chinese
Buddhist Temple provides Buddhist donations to turtle or bird rehabilitators,
supporting the latter's efforts to collect and rehabilitate injured animals. In
the past ten years, the Temple has donated over $5,000 every year, with a
total of about $80,000 since 2010. (2) After the fully recovered turtles and
tortoises reach a certain number (usually about ten), the rehabilitators will
call the Temple and make an appointment for the ritual release time. If the
Buddhists wish to initiate a "wise release of life," the Temple will contact
the rehabilitators and arrange an agreed upon date and place. (3) The rehabilitators link up with various nature reserves in the greater New York City
area, including Teatown Lake Reservation, Central Park in Manhattan, and
the Jamaica Bay Wildlife Refuge in Queens among others, to determine the
appropriate location and time for the release to ensure that the released animals can grow healthily without causing harm to themselves, other animals,
and the environment. (4) At the time of the release, Buddhists and the secular experts of animal rehabilitation and ecological protection together bring

Figure 6.1 Releasing life ritual for turtles in Central Park, July 19, 2018. Second to the left: Lorri Cramer; Third to the left: Abbess Jingyi. Photo by Wei Dedong.

the animals to the place of release. At the site, lay Buddhists and secular animal rehabilitation experts participate as the Buddhist monastics conduct a release ceremony for about one hour or less depending on the animals' ability to remain enclosed. After the animals take the "three refuges" of the Buddha, the Dharma, and the Sangha, they will be released and returned to nature.

II. THE BUDDHIST TEACHING OF "RELEASING LIFE"

The core of the Buddhist outlook on life is the concept of *karmic* response, which means good behaviors will be rewarded with good, and evil behaviors will be repaid with evil. Buddhists also believe there are six realms of rebirth in a cyclical process called *Samsara* (*lunhui* 輪回). The lowest realm is "hell," and above that is the "ghost" realm, followed by the realm of "animals." The three auspicious higher realms are the realms of "human beings," then "*asuras*," and the highest of the realms of rebirth is the "celestial" realm. Rebirth does not always go in a certain direction. After death, a sentient being's rebirth could go upward toward higher more desirable realms, or it

Figure 6.2 Chanting buddhist scripture during a releasing life ritual for turtles in Central Park, July 19, 2018. Far left: Lorri Cramer; Third to the left: Abbess Jingyi; Far right: Venerable Benkong. Photo by Wei Dedong.

could go downward, assuming less desirable life forms, or it can jump from one realm to another of the six realms. When people die, based on the *karma* that they created with their thoughts, speech, and actions during their lifetime, they will be reborn into one of these six realms.

There are many methods to accumulate merit, one of which is to release life. In the *Great Treatise on the Perfection of Wisdom* (*Mahā prajnāpāramitā śāstra*), the great Indian Buddhist philosopher and Bodhisattva Nāgārjuna (approx. 150–250 CE) said: "Among all the sins, killing is the first. Among all the merits, not killing is the first" (Nāgārjuna《大智度论》："諸餘罪中，殺罪最重；諸功德中，不殺第一"). However, in Chinese folk sayings, this was changed to, "Among all the sins, killing is the first. Among all the merits, releasing life is the first" ("諸餘罪中，殺罪最重；諸功德中，放生第一"). Not killing sentient beings is to protect life by observing the religious prohibition, while "releasing life" is protecting life and showing Buddhist compassion very proactively.

What are the specific merits of practicing the ritual of "releasing life"? In the enduring tradition of Chinese Buddhism, the lay public believes that there are two outstanding merits to be gained in conducting "release life" rituals: promoting one's longevity and good treatment by divinities. In the Buddhist ritual handbook, *"The Orbits of Releasing Life"* 《放生儀軌》, edited in the Qing Dynasty (Anonymous:28), it says, "If you want to prolong your life, you must release animals. This is the truth of the life cycle. When an animal is going to die, you must save it; when you are dying, then Heaven will save you" (汝欲延生須放生，此是循環真道理，他若死時你救他，你若死時天救你).

Why is "releasing life" such an important ritual that so many Buddhists feel compelled to perform? The reason greatly stems from Buddhist ontology and its basic concept of life. According to the Buddhist perspective, animals and people are basically equal; they are different categories of sentient beings who are reborn into one of the six realms of existence after death. For Buddhists, all sentient beings, including animals, ghosts, and celestial beings, have the potential of one day attaining Buddhahood and Enlightenment through their *karmic* actions. That means, saving an animal in this life is accumulating *karma* that will end up saving yourself or even saving a being who may one day become a Buddha.

Furthermore, since all sentient beings of any of the six realms of existence are endlessly recycled and reborn from one life to another, Buddhists believe that when you eat meat, you may be eating the flesh of your past or future parent. The *Brahmajāla Sūtra* 《梵網經》 says: "All the six realms of sentient beings are my parents. Thus, to kill animals and eat them, is to kill my own parents" (故六道眾生皆是我父母，而殺而食者，即殺我父母). The possibility and horror of unintentionally killing one's own parent is very real for Buddhists because time is infinite, and rebirths are also infinite. A human being or other sentient being is forever reborn within the six realms of rebirth until the being finally attains Enlightenment. During these unimaginably long cycles of time, every being will, at one time or another, be reborn into each and every one of the six realms. A human will be moving up and down the six realms, just like any other animal. In all likelihood, within the infinity of time, any animal may at some point, have been your parent, or will be your parent at some time in the future. This Buddhist doctrine about animals being our human parents at some point in our endless cycles of rebirth offers an important motivation for Buddhists to participate in "release life" rituals, even today, in our secularized modern society.

Unlike secular or non-Buddhist animal protection organizations, Buddhist "release life" rituals do not just set animals free. The most crucial step in the Buddhist ceremony is to help these animals to "seek refuge" in the Buddha, the Dharma, and the Sangha. The religious objective of the Buddhist ritual is to enable the animals to become a buddha. According to Buddhist teachings, to be classified as a sentient being is to be in possession of feelings and consciousness. Today's zoologists and biologists increasingly acknowledge that animals have both feelings and consciousness, although some scientists continue to use

animals for cruel and painful experimentations. Certainly, animal conservationists and animal rights activists promote the idea of animal feelings today,
pushing back against the modern history of Western science that produced a
nature that was inert and utterly different from human culture (Merchant 1992).
According to the concept of rebirth, if the animals can come upon human
beings teaching Buddhism, the Buddhist teachings will be planted into the
animal's consciousness, and become "seeds" that will influence the animal's
actions, creating karmic consequences that will steer the animal's future.

These animals that have been released in the Buddhist "release life" ceremony have gained three great blessings. First, they are no longer subject to
unfortunate events such as being captured or eaten, and can enjoy their lives
until they die of natural causes. Second, after dying, because of the blessings
of the "Three Treasures" (the Buddha, the Dharma, and the Sangha), these
animals will be reborn into the Western Pure Land or Western Paradise. Third,
these animals will practice the precepts in the Pure Land, meet the Buddha
there, hear His teachings, and become buddhas themselves to liberate other
beings from suffering. This is an important reason for Buddhists to perform
the "release life" ceremony, which is an expression of Buddhist compassion.

Due to the importance of the ritual of releasing animals, it is the project
that is the easiest to raise donations for Buddhist organizations. When people
go to Buddhist monasteries or temples, they are most happy to donate to
"release life" events. Whenever they or their loved ones become sick, or they
face a life crisis, people often make large donations to ask the buddhas and
bodhisattvas to help them.

III. THE "BLIND RELEASE OF LIFE" IN NEW YORK CITY

With the recent introduction of Chinese Buddhism into the United States, the
practice of "releasing life" has also arrived, and some of these rituals take the
form of what is called "blind release of life," which endangers the very life of
the animals that Buddhists are trying to liberate through the ritual. One reason
for this form of releasing in the United States is the lack of basic zoological
and ecological knowledge, as well as of the regulations concerning the treatment of animals and environmental protection in American law. Another
reason, however, is the basic economic interest, needs, and motivation of the
Buddhist temples and monasteries.

Venerable Benkong described the first blind release he attended during the
early days of his Buddhist service in New York City in 2006:

> I did not know that I was going to an animal release ritual. I didn't even know
> that there were such ceremonies when I began to reside at the Grace Gratitude
> Buddhist Temple in Manhattan's Chinatown in 2005. What the abbess of the

temple told me was that a member of the congregation, who lived on Staten Island, close to the shore, had an elderly dog that was soon going to die. She hoped that this dog would be reborn into the human realm where it would be able to come into contact with Buddhist teachings and begin its path to Enlightenment.

We all met at the home of the Buddhist dog-owner on Staten Island. Her dog was present, lying quietly on a duvet in the living room. One by one, each person went up to the dog to pray, and the Buddhist monks blessed the dog. Then, after we were all treated to snacks, our caravan of cars drove to a nearby beach. As the members of the congregation began showing up and parking near the shore, two vehicles off-loaded two very large baskets from the trunks of their cars. The baskets were filled with crabs, which had been in the trunks of their cars for hours. I didn't immediately understand what the crabs were there for, because it was in the dead of winter. I never thought they were to be released into the freezing cold water of New York Harbor!

The abbess ordered the congregants to put on their robes and gather around the baskets of crabs. Then the ceremony began with Buddhist scripture chanting, reciting, and playing our musical instruments for about an hour. The cold wind blew fiercely on the poor exposed crabs that had been in the car trunks for hours, during the 50-mile drive from Manhattan Chinatown to Staten Island during bumper to bumper rush hour traffic, while we were enjoying our snacks in the warm interior of the congregant's home.

The neighbors of the woman dog-owner started to come out of their houses to see what we were doing. They were as shocked as I was when the male congregants began carrying the baskets of crabs onto the sand at the edge of the surf, and undoing the wires that held them shut. I was ignored by the abbess and the twenty or so congregants when I cried out, "Wait a minute, are those crabs indigenous to this area? How do you know they can survive here? Why are you letting so many of them go in the same place?" The men opened the baskets and dumped the hundreds of crabs onto the cold wet sand. As the waves rolled over the crabs, several neighbors were now gathered on the beach asking, "What are you people doing?!"

As the crabs hit the freezing water, they were unable to move. They just laid there, some on their stomachs, some on their backs. Sea gulls are everywhere in Staten Island, as common as pigeons in Manhattan. When the first sea gull arrived, it pierced a crab that was laying on its back through its chest, hooking it onto its bill. As soon as it flew away with the crab, other sea gulls came one by one. Soon, there must have been a hundred sea gulls circling overhead and picking at the crabs that lay strewn all along the beach by the waves. The crabs were being knocked back and forth from the sand to the sea by the surf. I didn't see a single crab scurry away or hide itself in the sand.

As the neighbors gathered in shock and awe, the congregation ignored their comments as they happily got into their vehicles, proud of the success of their "blind release." We returned to the house, gathered around the old dog, and transferred the merit of this release to the dog. Then the dog-owner gave each of us monastics a red envelope of cash to thank us for our compassionate work.

Everyone was sure that we had guaranteed the rebirth of the dog into the human realm. The owner invited all of us to a local restaurant for a vegetarian dinner. And the merit of this offering was also transferred to her dog. A waitress, obviously unaccustomed to seeing a group of Buddhist monastics, asked me, "What brings you all to Staten Island?" I answered, "We came to feed the sea gulls."

The "blind release of life" carried out by Chinese Buddhists in the New York City area even caused some judicial intervention. In 2004, a laywoman's mother got cancer and she underwent chemotherapy. The laywoman bought 500 small turtles and followed the Abbess Jingyi to release them into the seawater of Boston Harbor to accumulate merit to be transferred to her mother. Unfortunately, these were freshwater turtles and naturally, they all eventually perished there. An American journalist accompanied them, and wrote up a news story about the Buddhist ritual for a local newspaper, along with an enlarged photo of the turtles being dropped into the sea. The news story greatly damaged the reputation of Buddhism. The journalist did not expect to cause such a negative uproar. When the Boston locals saw that so many turtles were suddenly released into the water, they became concerned about the ecological balance and the survival of the turtles, and many alerted the local authorities. Abbess Jingyi faced a fine of $50,000 and also possible penalties of imprisonment. She was greatly shocked. In a stroke of good fortune, the presiding judge unexpectedly knew something about Buddhist culture, and understood the meaning of Buddhist rituals of "releasing life." The judge reduced the punishment, seeing that the Chinese Buddhists were first offenders, who did not understand animal survival, but sincerely believed they were doing a good deed. However, he lectured Abbess Jingyi that she cannot just casually release large numbers of animals in the United States without permits.

In addition to the lack of knowledge, another reason for the phenomenon of "blind release" in the United States is the Buddhist temple or monastery's economic interests. Generally, in New York City Chinatown, a temple charges each participant of a "releasing life" ritual about $50. The temple would then charter a sixty-five-seat bus for the trip to the release site, and would provide box lunches prepared by temple volunteers and staff, for the participants. The temple would even buy baskets of turtles from Chinatown butcher shops to release. The temple's investment in each participant was probably $20 per person, thus it was making $30 on each person attending. This amount provides for the upkeep of the temple.

It is no surprise that the "blind release of life" has damaged the reputation of Buddhism in the United States. The American media has been criticizing the practice since the 1990s. One article in *Scientific American* wrote, "Buddhists across Asia release wildlife as a show of compassion, but conservationists find that the practice tortures the animals and may impact threatened species" (Nuwer 2012). *The New York Times* featured an article that acknowledged that Buddhists like to accumulate good seeds, "But merit is harder to come by in these times . . . . The person releasing the turtles may have meant well, but it was fatal to these creatures" (Westjan 1997).

IV. THE SYNCRETIC INNOVATION OF "WISE RELEASE OF LIFE" IN NEW YORK CITY

The "wise release of life" illustrates Chinese philosopher Zhang Dainian's Theory of Syncretic Innovation at work in Chinese Buddhist practice in New York City Chinatown. It combines the Buddhist notion of compassion and merit in saving sentient beings, with modern scientific approaches to animal rehabilitation and maintaining ecological balance. Thus, "wise release" realizes the creative transformation of traditional Buddhist culture adapted to a modern urban context. It insists on practicing ancient Buddhist rituals and continuing the age-old tradition of accumulating merit, while avoiding the environmental destructiveness of unreflective practices of "blind release."

One of the first large-scale-wise release rituals was held on August of 2010 at the Teatown Lake Reservation, about thirty-two miles north of Manhattan. About thirty members of GGBT joined several Teatown Lake Reservation staff for this event. Teatown was an ideal location because of its facilities specifically designed to provide education about the environment. Both students and visitors are welcomed. Teatown also provides onsite rehabilitation for various kinds of animals. The Buddhists were able to use their facility to perform the release ceremony.

At the beginning, an animal rehabilitator and educator at Teatown gave a slide show presentation to introduce the stories of each turtle who would be released. What caused the turtles and tortoises to be in the care of the rehabilitators? There were photos of the injured animals, the treatment strategies and prognoses were explained, as well as the length of time and costs incurred until the animals were now well enough to be released. The Chinese Buddhist participants were in awe that such people would care for a turtle for two years, changing bandages every couple of days. Some turtles also carried trackers and could be positioned by GPS to keep abreast of the turtles' activities. These compassionate actions greatly moved Abbess Jingyi and the other Buddhists.

To the embarrassment of the Buddhists present, some of the turtles were obviously the victims of previous "blind releasing," because they were discovered frozen in lake water, or dying in salt water, even though they were freshwater turtles. The rehabilitators explained how much suffering these

animals experienced because they were released in the wrong location or at
the wrong time. Freshwater turtles released into salt water will absorb salt,
which their bodies cannot purge, so they die a slow death as their livers and
spleens burst over several weeks. Southern turtles cannot hibernate, and they
wind up freezing to death in the winter of northern climates. The first organs
to freeze are their eyes, with the water in the eye cells crystallizing and causing them to go blind painfully.

After the education program, the Buddhists of Grace Gratitude Temple
performed the Buddhist "release life" ceremony. Every Buddhist receives
the ritual handbook, *"The Orbits of Releasing Life"*《放生儀軌》, which
consists of three main sections: (1) Buddhist sutras and mantras to be chanted
during the ritual, for purifying the environment and praying to the buddhas
and bodhisattvas to appear and support the ceremony; 2) monastics explain-
ing the meaning of releasing animals and enabling them to seek the three refuges; and (3) the ritual steps that Buddhists take to release the animals. These
three sections are further divided into seventeen smaller steps or procedures,
which will be detailed later.

As the Buddhists performed the ceremony, Venerable Benkong explained
in English to the assembly:

> We are performing the ceremony in the hopes that these animals will not be
> abused, will not be caught or trapped, and will not be murdered, so that they
> could live out their lives peacefully and naturally. We also hope that one day,
> they may develop compassion and knowledge, and eventually be reborn out of
> the animal realm. We hope that one day they can become human, will hear the
> Dharma, have an affinity with it and hopefully accept it. If they practice it, they
> will become buddhas themselves too, and in turn, be able to liberate others from
> suffering.

The animal rehabilitators who were present had never taken part in a release
ceremony, nor had they ever participated in any Buddhist ritual before. This
was the first time that they had ever witnessed "a religious ceremony of any
religion that specifically focused on the well-being and liberation of animals,"
wrote Venerable Benkong in response to a question about what some of the
animal rehabilitators learned from their cooperation with Chinese Buddhists.
Some of the staff were young, novice, rehabilitation interns, and they were
either college students or recent graduates. For them, Buddhism was an alien
and exotic religion, about which they knew very little. Venerable Benkong
made two observations about the impact of the Chinese Buddhism upon the
rehabilitators. First, the animal rehabilitators were surprised to learn that
"Buddhists did not see animals as forever animals, but saw the turtles as
future Buddhas." Second, they witnessed a concrete demonstration of how
the notion of *karma* works, that for Buddhists, the act of releasing animals

and donating thousands of dollars to the animal rehabilitation organizations, would produce "positive *karma* and even mitigate negative *karma*." Thus, the secular work of environmental and animal protection and conservation are greatly enriched by Buddhist ontology. They had been introduced to a new ontological orientation that saw humans and animal species as having inter-linked lives, and was not merely concerned with the short biological lifespans of humans and animals but also with their many future interconnected lives of suffering and spiritual attainment.

As the ceremony proceeded, and while Venerable Benkong was expressing the hope that the released turtles and tortoises would not be harmed, or caught in traps and nets, and so on, tears welled up in the eyes of two young staff members at Teatown who were doing their internships. Just as the Chinese Buddhists, most of whom were immigrants, were so moved by the compas-sionate care that the American animal rehabilitators provided to the animals, the young American interns began quietly crying because they were moved by the Chinese Buddhist ritual. After the ceremony, Abbess Jingyi was so touched by their tears that she came to ask them why they were crying. They replied that they had never seen people exhibit so much love and care for the welfare of animals as shown in the Buddhist "release life" ritual.

In the last step of the ceremony, the animal rehabilitators, the Buddhists, and the Teatown staff took the animals to the individual places of release. This step of "wise releasing" contrasted with the "blind releasing" approach, which carelessly dumped the animals all in one spot without regard to where the animals originated. To the best of their ability, the rehabilitators brought the animals back to the location where they had been found injured, which was their own natural habitat. Others animals, like the turtle hatchlings which came out of the eggs of dead turtles that had been incubated by Teatown staff, were released around the lake or pond near where their mother had died. These baby turtles had to be released a distance of at least ten feet from each other, so that there would be enough food for them to eat.

Before traveling to the Teatown Lake Reservation, the Abbess Jingyi had asked the rehabilitators what the optimum amount would be to cover the med-ical costs of one turtle, and learned that it would normally be $20. The abbess had informed her Buddhist congregation of this amount, but left it up to them to donate what they felt would be suitable. The Buddhists were all so moved by the presentation, the care and dedication by the Teatown staff and reha-bilitators, that they each donated more than the basic sum. The temple cut a check matching the total donation of all the Buddhist participants combined, with all the donations going to the Teatown Lake Reservation and rehabilita-tors. Thus, in this evolving "wise release" approach, Chinese Buddhists in the United States have also begun to donate money to mainstream and secular animal protection organizations in the United States to treat injured animals.

Releasing animals in cooperation with different animal and bird protection organizations is gradually becoming an approach that Chinese Buddhists now accept as a better way to accumulate merit than buying animals for release. This marks a historic moment in the more than 2,500 years of Buddhist "releasing life" rituals: the birth of a new "Wisdom of Releasing Life"!

The Ritual Steps in the Wise Release of Life

As mentioned earlier, the "wise release of life" ceremony consists of seventeen steps. These are based on the Chinese Buddhist ritual handbook, known as the "*Orbits of Releasing Life*" 《放生儀軌》. Venerable Benkong translated it into English and printed a brief bilingual edition for Buddhists to use. The entire ceremony with all seventeen steps takes about an hour to perform. However, since the "wise" aspects of the ceremony is in tension with the well-being of the animals to be released, the ceremony is usually shortened, depending on the animals' endurance and stress level. The "wise release of life" ritual steps follow the order given next (table 6.1):

Extending the Range of Wise Release from Animals to Birds

During the autumn of 2014, a journalist asked the GGBT to do a wise release for a story she was writing for the *Audubon Magazine*. Venerable Benkong explained to her that they could only release animals in the summer because turtles and tortoises are cold blooded. However, the journalist suggested that they could get in touch with the WBF in Manhattan because they release birds almost daily into Central Park. For this reason, the "wise release of life" came to be extended to the release of birds into the wild.

New York City is a major stopover on the East Coast migratory flyway, and over 355 bird species live in the Big Apple, or take refuge there during the spring and fall migrations. There are many skyscrapers in Manhattan, so many birds are injured by flying into buildings that have sprung up along these ancient migratory pathways. The many cats in Manhattan who attack birds, vehicles, and human interference, such as kids with BB Guns, also take their toll on birds.

The WBF is a state and federally licensed 501(c)(3) organization, whose mission is to provide medical care and rehabilitation to native and passing migratory wildlife, so that they can be released back into the wild. The work of WBF's bird rehabilitation includes the actions of diagnostic testing, surgery, applying medication, bandaging, splinting, physical therapy, feeding, sheltering, and the use of radiographs. All native and migratory birds are treated, from house sparrows to rare species such as Virginia rails and Great-Horned Owls. WBF is conveniently located on the edge of the Central Park, so when

Table 6.1   Ritual Steps of the Buddhist "Wise Releasing of Life" Ritual (reconstructed from The Orbits of Releasing Life 《放生儀軌》. Anonymous 1980)

	Setting Up the Ceremony 準備	An incense table is set up at the ritual site.
1	Honoring with Incense 香贊	The buddhas and bodhisattvas are welcomed to come and attend the ceremony.
2	Calling the Sacred One 稱聖號	The name of the Compassionate Bodhisattva Guan Yin is invoked.
3	The Master Speaks over the Sacred Water 法師說水文	The presiding Buddhist monk or nun purifies the ritual site with sacred water.
4	The Great Compassion Mantra 大悲咒	The presiding monk or nun and Buddhist participants repeatedly recite the Great Compassion Mantra as sacred water is sprinkled, purifying the environment.
5	The Heart Sutra 般若波羅蜜多心經	The Heart Sutra is recited to embed it into the minds of all present, including the animals, that all may realize sunyata (emptiness), release themselves from excessive desires, and be liberated from suffering.
6	The Rebirth Mantra 往生咒	This mantra is chanted to instill in the minds of the animals the idea that they may be reborn into a higher realm of consciousness.
7	Inviting the Three Treasures 請聖	This step is to invite buddhas, the Dharma (universal truth), and the Sangha (monks and nuns) to be present.
8	Words from the Master 法師白雲	The presiding monk or nun explains to the animals and participants that through this ceremony, all are on the Bodhisattva Path, and sentient beings will be released. The ritual leader expresses the wish that all will attain Enlightenment.
9	Repentance 懺悔	All present repent for their negative karma that was created out of ignorance, anger, and greed, and all vow not to create it again.
10	Mantra Recitation 持咒	All chant mantras to strengthen their vows.
11	Transferring Refuge in the Three Treasures 傳授皈依	The animals and human participants now "take refuge in the Three Treasures" to practice becoming buddhas, by learning the Dharma with the support of monks and nuns.
12	The Four Great Bodhisattva Vows 發願	All vow to liberate all sentient beings from suffering, to end our delusions, to learn the Dharma, and to become buddhas.
13	Praising the Names of the Tathāgata 稱讚如來名號	By praising the name of the Tathāgata, the participants can become true disciples of the buddhas.

(Continued)

Table 6.1   Ritual Steps of the Buddhist "Wise Releasing of Life" Ritual (reconstructed from The Orbits of Releasing Life 《放生儀軌》**. Anonymous 1980)** (*Continued*)

	Setting Up the Ceremony 準備	An incense table is set up at the ritual site.
14	Releasing the Animals 放生	Before the animals are released, participants make a wish that they will not suffer, nor create more negative *karma*. They wish that they may live well and be reborn into a Pure Land where they learn to become buddhas.
15	Praising the Animals 唱讚	Participants wish that the animals may "take refuge in the Three Treasures" and devote themselves to self-liberation and liberating others, which will help them live free and at ease, and be reborn into a Pure Land.
16	Chanting the Names of buddhas and bodhisattvas 念佛	As the animals are released back into nature, participants chant the names of buddhas and bodhisattvas to wish that they will protect and help these animals in the wild. They wish they may be reborn into the human realm one day, come in contact with Buddhism, develop themselves, and be reborn into the Pure Land.
17	Transferring Merit 迴向	Participants transfer the merit of this Wise Releasing of Life Ceremony to all participants who have rehabilitated the animals and have offered support. They wish that they will be released from suffering and will improve themselves and help others.

people find an injured bird, they can contact its office directly, or find it through the New York City Police Department. The founder and head of WBF welcomed the idea of bringing Buddhist participants to their tiny bird rehabilitation center near Central Park. It is also a short subway ride from Chinatown, so the GGBT congregants could leave at any time for home on their own.

On Sunday, October 23 in 2016, six Buddhist monastics and forty lay Buddhist members participated in the wise release of birds in Central Park. At 2 o'clock in the afternoon, the Buddhists gathered at the GGBT in Chinatown and took the subway under the leadership of Abbess Shi Jingyi. After less than an hour, they arrived at the subway station nearby the WBF. Under the guidance of professionals, they admired the lovely birds who were to be released that day. Each bird was placed in a special box or cage. The pigeons were held in paper cartons that had several holes each. The other birds were in iron cages that were tagged with different colors of cloth. The WBF has professional regulations on how the birds are dressed for injuries, and the length of time for recovery before they can be released into Central Park. For example, the small birds must not be released at the active dusk or night time of the owls, to avoid their being eaten, and so forth. That day, a total of nine birds, including five pigeons, a wild duck, and three wild birds were released. Following the bird experts, the Buddhist ritual participants carried a variety of bird boxes and cages onto the Great Lawn in Central Park (figure 6.3).

The Abbess Jingyi led the group through the steps of "wise release" ceremony step by step. After she finished the chanting and invocation of the buddhas, the Buddhists opened the boxes and cages and released the birds into the clear sky. One of the birds was reluctant to fly off. Another bird was taken alone to a pool by a WBF bird expert, who said that the bird needed to be released near water. Indeed, only the rehabilitation specialists were able to know the animal's needs so accurately, and to make the release based on scientific knowledge.

In 2016, I published a report on this wise release of birds in Central Park in *China Ethnic News*, a Chinese newspaper about ethnic minority affairs in China (Wei 2016) and he received a flurry of warm responses from the Chinese Buddhist readership. This is the first time that Chinese readers learned of the "wise release of life" from an authoritative newspaper. It is clear that Chinese readers appreciated this news, because the "blind release of life" is so common in Mainland China, and has become a social crisis today.

Cooperation with the Environmental Protection Agency

With a growing reputation in New York for the "wise release of life," the GGBT soon worked directly with the New York City government to help it solve some social problems. These cooperative activities have gained a positive reputation for Chinese Buddhists in New York City. On May 2015, officers from the Environmental Protection Agency (EPA) contacted Ms. Lorri Cramer of the NYTTS, saying that they had confiscated a few hundred baby red-ear slider turtles. The officers were not sure from where the turtles originated. They may have come from a criminal ring that smuggled them from the South or the Gulf of Mexico area, across state lines to New York City's illegal pet market. They could also have been raised on turtle farms in China and smuggled in from overseas. The reason for the China connection was because the containers they were found in had Chinese characters on them.

The two officers said that the turtles were confiscated because baby red-ear sliders are not permitted to be sold. The reason is that, under the age of eight, they often carry salmonella. Although not permitted to be sold in the pet industry, many street vendors in Chinatown sell little red-eared sliders together with the plastic container they are trapped in, with a little bit of water for only around $5. It is a very hot item of purchase for tourists who frequent

Figure 6.3   Releasing life ritual for birds in Central Park, Oct. 23, 2016. Front left: author Wei Dedong carrying bird box provided by the Wild Bird Fund. Photo by passerby.

Chinatown and do not understand that the turtles carry salmonella, and that children especially should avoid coming in contact with them. The officers said that they would normally kill the couple hundred of the turtles by freezing them to death. Before taking such a drastic measure, they hoped that Lorri Cramer might know of a place that was willing to take them in. Normally, Lorri would send abandoned red-ear slider turtles to the South, where they are indigenous, to a state that does not have laws against importing them. However, since these turtles might be carrying some kind of virus, they could only be released into an enclosed pond, and not released into the wild, where they may infect local populations with an infectious antigen that they may be carrying. So Lorri took in the couple of hundred turtles, not realizing that the officers would soon make another bust of a couple of hundred more turtles, and then another. By the time the total population of smuggled and confiscated turtles were counted up, there were a total of 625 baby red-eared sliders in all! The babies were so stressed by being packed in thick layers of turtles with very little water, and they obviously had not been fed or cleaned. Furthermore, the turtles at the bottom of the containers had either drowned or been crushed, so the survivors were reduced to only 570 turtles left.

Lorri thought to call Venerable Benkong and asked for help from Buddhists. Venerable Benkong suggested that the Buddhists first bless the turtles with positive energy so that they could find homes. Lorri's own apartment building manager agreed that GGBT could have the use of the rooftop on her apartment building for the blessing ceremony. The GGBT then contacted the American Buddhist Confederation, requesting that they send volunteers to come help move so many cartons of turtles to the roof. However, before they began the blessing ceremony, the local media showed up to document the event. There were newspaper journalists, as well as television people with lots of television equipment, as well as people who worked for radio stations. The poor turtles were already stressed out as it was, and now there were all these media people adding to the chaos. Lorri was very upset and worried that her apartment manager might be very angry that she had invited the media without first informing him.

Fortunately, there is usually a silver lining to every crisis. The media descent and the hype that they created did also publicize our need for people or organizations to adopt these turtles and prevent them from being killed. Many generous people and organizations stepped forward, and they were given instructions, "What you Need to Know" and "Do's and Don'ts" in caring for red-eared slider turtles. All were informed about how to avoid being infected with salmonella, and warned to keep the turtles out of contact with children. Everyone was informed that the turtles would not be released into the wild. The people who agreed to adopt were also informed that these turtles can live up to eighty-five years, so taking in a turtle as a pet is a not only a lifetime commitment, but could become a multi-generational commitment. Furthermore, they found out that the turtles could grow to the size of a dinner plate, so they needed a very large living environment with both land and water, and a basking area under the sun to maintain their health and happiness.

Conflicts between Wise Release and Blind Release Life

The "wise release of life" reflects the syncretic innovation of traditional Buddhism that is unprecedented in the history of Chinese Buddhism. In actual practice, the wise release inevitably conflicts with traditional approaches to releasing, especially with "blind releasing." The popularity of this ritual is due to a large extent, to the temple or monastic economy. Many temples rely on releasing rituals to raise funds, and the transition from "blind releasing" to "wise releasing" requires a process that is not accomplished overnight.

After the news that Grace Gratitude Temple cooperated with the NYTTS to carry out releasing rituals in the August of 2010, it immediately received the attention of many social groups, including the International Humane Society (IHS) in Washington, D.C. After conducting a written interview with Venerable Benkong and coming to understand the basic facts about this Buddhist ritual, an IHS team traveled up from Washington, D.C. to New York City at the end of December of 2010. They had in mind the promotion of "wise releasing" among the members of the American Buddhist Confederation (ABC). The IHS had already established a program to educate Buddhist temples in the United States about the harm that the "blind release of life" caused to the environment, and that it was cruel to animals. It put on a presentation at an ABC gathering in Manhattan, where there were roughly twenty ABC Buddhist temples from the tristate area represented. The IHS showed the Buddhist temples a film about the harm that "blind animal releases" cause in China, where many animals to be released died in their cages during transport. The film, which was in Mandarin Chinese, also showed how after the release, many animals wound up dying in the vicinity of the release. It presented graphic images of animal corpses strewn all over the ground the very next day after a release, because they could find no food, water, or shelter in the alien territory. Confused by the new environment into which they were dumped, they could not find any refuge or protection. Some birds died overnight because of the nighttime drop in temperature, or they could not find appropriate trees or holes in rocks to nest.

In their compassion, all the ABC member monks and nuns shook their heads and made a "tsk! tsk!" sound at the horror of animal suffering. Everyone agreed that current practices were often very harmful and irrational, and not at all compassionate. Some of these monastics were obviously embarrassed. IHS took a very direct approach by going around the table and having each person state that they now understood how harmful was the carelessness of "blind release," and that they would never do another blind release. At last, all the representatives raised their hands as if they were swearing an oath in court, and stated, "I will never engage in the blind release of animals again. It is very harmful and cruel."

The IHS contingent were very happy at the success of their program. They were relieved that it was so easy to "reeducate" these seemingly agreeable, non-argumentative, non-defensive Chinese monastics. The head of the IHS contingent thanked everyone for coming, especially the leader of ABC and Venerable Benkong.

Much to people's surprise and disappointment, at the next American Buddhist Confederation meeting two weeks later, a Buddhist nun and her temple had a "blind release" planned for the date of the next ABC meeting and so would be unable to attend. Venerable Benkong was stunned to hear this, and angrily reminded her that two weeks prior, the abbot of her temple had raised his hand and sworn in front of the head of IHS that they would never do a blind release again. She defended this release by saying that it had been planned before the IHS showed their film. She argued they had already chartered a sixty-five-seat bus, made a purchase order of sixty-five turtles, and people had already given donations for the lunch boxes. She insisted that they had been collecting $45 per person for the event, and it was too late to cancel it.

Venerable Benkong also stood his ground, warning that if someone at this releasing event were to take some photographs, and if the International Humane Society should get wind of this serious lapse of judgment by a Buddhist temple, then IHS would lose all faith and trust in the ABC, and accuse the organization and its temple members of being liars. Performing a "blind release" only two weeks after all the monastics at every temple swore not to do it the old way makes the entire Chinese Buddhist community look pretty bad, Venerable Benkong pointed out. This raised the ethnic or nationalist hackles of one Chinese monk, who heatedly demanded that Master Benkong, an American, answer his question, "How many people is your American war in Iraq killing? You say nothing about that! All you care about is the welfare of some turtles!" This escalated the tension and further angered Master Benkong. He explained that he had been invited to join Grace Gratitude Temple and ABC to educate the Chinese Buddhist community about what is appropriate and inappropriate in America, and also to inform people about the laws and regulations in this country. "Animal welfare is serious in the U.S.," he declared, "The laws of New York State are clear about releasing animals into the wild: it is illegal! Why are the temples guiding their congregations to break the law?" Finally, he declared that he could no longer be part of ABC, so he resigned his position as a consultant.

The aforementioned example, where there were genuine good intentions to start practicing "wise release," but often there is a reversion to the old familiar practices of "blind release," is fairly typical among the Chinese Buddhist community in New York City. I suggest a few possible reasons (or hypotheses) for the reversion to "blind release," even in the face of newly acquired knowledge about the suffering and death it inflicts upon the animals that are supposed to be saved. *First*, in Asia, and now among many Chinese Buddhist temples in the United States, temples have an economic interest to organize many "blind release" ritual events because they can bring in lay Buddhist donations, or funding for temple upkeep and expansion. In contrast, "wise release" occurs less frequently because it must be coordinated with various animal welfare agencies according to their needs. Moreover, "wise release" does not produce any income for temples because it involves the temples donating money to these agencies instead. *Second*, in recent decades, China has gone through radical unprecedented industrialization and urbanization drives, and many people have lost touch with their rural agricultural roots, where people understood better the biological and environmental requirements for the survival of animals, fish, and birds. The *third* possible explanation, however, partially negates the second: ordinary people do not have much scientific education, so they often do not understand some basics of nonhuman species diversity and biological survival. *Fourth*, while Buddhist sutras possess strong threads of compassion for all sentient beings, including nonhuman beings, perhaps in actual practice down through the ages, human beings have always privileged themselves. This basic anthropocentrism has been exacerbated by modern times, with both Western liberal humanism and Chinese revolutionary humanism, and the penetration of capitalism's production and consumerism. This fourth explanation is one based on a deeper history. I present these four possibilities, not as final explanations, but to spur further research about the history and present situation of the Buddhist "release life" rituals.

The Limits of Syncretic Innovation and the Difficulties of Wise Release

The "wise release of life" is an example of what Zhang Dainian called a "syncretic innovation" in New York Chinese Buddhist practice. However, this path of syncretism and hybridity is a difficult one, and new tensions, negotiations, and obstacles are continuously unfolding. In the collaboration between Grace Gratitude Temple with the Jamaica Bay Wildlife Refuge, I discovered the hard limits of their innovation. They were able, even eager to release animals in cooperation with the Wildlife Refuge, however, they refused to donate to an important excavation project initiated by the Refuge that would protect both the wildlife and the environment.

The Jamaica Bay Wildlife Refuge, part of the Gateway National Recreation Area to the southeastern side of Brooklyn, is one of the most important urban wildlife refuges in the United States. It is made up of a salt marsh, woods, and freshwater ponds, located on 9,155 breathtakingly scenic acres, an oasis in the middle of the greater New York area.[2] Like Teatown Lake, the Jamaica Bay Wildlife Refuge has a Visitor Center to educate people about their famous native resident, the Diamondback Terrapin,[3] a turtle that lives in brackish water, or water that is a mixture of salty and fresh. These terrapins are also being captured by the pet industry because the dark outlines of their diamond-shaped designs on their shells are so unique and attractive that they have both a domestic American and an Asian market. They are even being bred in Japan, where their shells can be morphed into many different colors.

The terrapin plays a crucial role in the health and resilience of the Refuge's ecosystem, but indiscriminate commercial crab traps, habitat loss, and polluted waters in Jamaica Bay continue to threaten their existence. The terrapin population in Jamaica Bay has declined by more than one-half in the past decade because of a falling egg-laying capacity in female terrapins. The Diamondbacks feed in the Bay, primarily on periwinkle snails. The Periwinkles feed on salt marsh cordgrass, a primary vegetation of the Bay's marshland. In short: without terrapins, the periwinkle population would grow unchecked and decimate the grasses that make up the bulk of the marsh. Without Diamondbacks, the marshes—which are already eroding—could decline further. This, in turn, would further erode the Bay's resilience and ability to serve as a buffer for New York City against extreme weather events, such as Hurricane Sandy in 2012.

Jamaica Bay has a Diamondback breeding program where wildlife conservationists and other staff go in search of their nests and surround the nests with wire mesh, so that rats, raccoons and other predators would not destroy the eggs. Whenever the staff saw the ground moving, they would wait for the terrapin hatchlings to crawl out and would stay to protect them from predators. When all the hatchlings emerged, they would remove the mesh and allow the baby terrapins to enter into the water, where they would continue to make efforts to protect them in the water. Thus, in order to prevent their extinction, the Diamondbacks required a great deal of attention in their natural environment.

Lorri Cramer introduced Venerable Benkong to a conservationist working with Diamondback terrapins at Jamaica Bay. Venerable Benkong invited the Buddhist Youth Communication group to participate in this program. The members of this Buddhist Youth organization were youth when they created this organization about forty years ago, but now most of them had become elderly people. Venerable Benkong said to them:

> We have a chance to move into a new approach to animal releasing with Jamaica Bay. Traditional release ceremonies rescued and released animals that were to be slaughtered for food. With Teatown, we evolved this into rescuing animals that were injured or that may die for whatever reason and releasing them after rehabilitators believed they were healthy enough to go back into the environment. From the original Bodhisattva vow of "seeing murder, go to the rescue" (見殺要救), now we are evolving into the perspective of "seeing death, go to the rescue" (見死要救). The Jamaica Bay concept is a new move into animal conservation.

Fortunately, many Chinese Buddhists were open to this new way of practicing the "release of life," and they were welcomed by the Jamaica Bay staff, who even agreed to hold back some hatchlings that were not strong enough to go into the water yet. As usual, the Buddhist members showed up, attended an education program, and were very happy to help release the baby terrapins into the water. As for those terrapins that had been held back, the Buddhists enjoyed holding them in their hands, and then slowly placing them into the water. Again, the Buddhists collected donations among themselves and transferred the merit accrued from this release to the animals, that they may be safe and attain Buddhahood. They also ensured that some of the merit would also go to congregants who made donations and to the Jamaica Bay Refuge, so that it may continue to be a sanctuary for wildlife. All went in accordance with the ancient Buddhist traditions of the ritual of "releasing life."

Unfortunately, the encounter of two very different cultures and sets of expectations do not always end with a positive outcome. The cooperative project between Chinese Buddhists and the Jamaica Refuge did not continue. The Jamaica Bay leaders and staff, seeing that the lay Buddhists were happy to support them in their good work, then requested that they donate to support an excavation project that would fix the damage caused to the small wildlife sanctuary when Hurricane Sandy brought devastation. Being elderly and on fixed income, the Buddhists were more interested in the traditional release. They loved bringing little turtles to the water. They understood the construction and excavation for conservation work needed to be done, but they were not willing to make donations for this new project, which did not seem to directly lead to merit accumulation as they understood it. Thus, issues of species extinction, ecological balance, preservation of habitat, and repair of marshland to build a buffer from hurricanes to protect a large urban population were too unfamiliar and distant from the task of ensuring a good rebirth for themselves and the terrapins, and the attainment of Buddhahood for all. For the time being, the Chinese Buddhists in New York City seem to have reached the limits of their Buddhist innovative capacity. However, things may change among the next generation of Buddhists.

V. CONCLUSION

The "wise release of life" is the successful syncretic innovation of traditional Asian Buddhism in the United States. The adaptation and syncretism of this ancient Buddhist tradition in the new context of contemporary American society harbors multiple meanings and significances. "Wise release" preserves the Asian Buddhist traditions of accumulating merit and meeting the religious needs of the immigrant Buddhists. From a sociology of religious perspective, in modern societies, those religious organizations that are more open to new membership and religious change may sometimes lose their believers; while those organizations that adopt a more conservative attitude toward religious doctrines and practices may have greater social cohesion. There are about fifty Chinese Buddhist temples in New York City. The majority of lay Buddhists are immigrants from the coastal Province of Fujian in

China. Most of them come from the lower middle class, and they have more traditional religious needs, such as the desire for a long life, good health and healing, honoring the memory of ancestors and kin, and the desire for good fortune and prosperity.

The "wise release of life" established the integration of traditional Buddhist notions of merit, modern animal rehabilitation, and environmental protection, reflecting the ability of Buddhism to adapt to modern society, science, and globalization. Buddhism is an ancient wisdom born in Asia, while the United States is a country of Christian and Jewish traditions. So the question arises of how Buddhism could take root in the United States and serve Americans in this era of scientific and technological innovation? This is a topic that all Buddhist groups must explore. I believe that the examples of the "wise release of life" presented above can help people rethink the practice of religion in modern secularized societies, and incorporate ancient alternative ontologies to carve out new paths forward for Buddhist contributions to environmentalist efforts.

The "wise release of life" offers an effective channel for mainstream American society and the public to accurately understand Buddhism. The United States deserves its good reputation and can be very proud of its numerous animal protection organizations and nature reserves, along with its national and state parks. Through the "wise release" rituals, these organizations in New York City have come to understand the teachings and practices of Buddhism. The reports of this practice on these organizations' websites, magazines, and newsletters, and in the mass media in general, have also greatly improved the image of Buddhism and contributed to the localization and embedding of Buddhism in the United States. Despite Buddhism's almost two-thousand-year history in China, its admittedly still brief settlement in the United States has already begun to change it, and these changes will likely be introduced back to China. In Buddhism's adaptation to American society and culture, the innovation of "wise release" may inspire new varieties of Buddhist environmental rituals and actions back in Asia and China.

NOTES

1. I am very grateful to Mayfair Yang for her contributions to the editing and polishing of this chapter.

2. https://www.nycgo.com/venues/jamaica-bay-wildlife-refuge (Accessed Sept. 16, 2020).

3. https://news.mongabay.com/2017/02/saving-jamaica-bays-diamondback-terrapins/ (Accessed Sept. 16, 2020).

CHINESE BIBLIOGRAPHY

Anonymous 佚名. 1980. *"The Orbits of Releasing Life"* 《放生儀軌》. Hongkong: Hong Kong Buddhist Book Distributor.

Dainian, Zhang and Cheng Yishan 張岱年、程宜山著. 2015. *The Spirit of Chinese Culture* 《中國文化精神》. Beijing: Peking University Press 北京大學出版社.

Fan Wang Jing 《梵網經》 (*Brahmajāla Sūtra*), in《日本大正新修大藏經》 (*Taisho Tripitaka*), Volume 24, page 1006b. *Chinese Buddhist Electronic Text Association* (CBETA) http://www.cbeta.org/download/ebook.php (Accessed Nov. 28, 2020).

Lian, Chi 蓮池. 2012. *"Ahimsa and Releasing Life Texts"* ("戒杀放生文") in *The Complete Works of Master Lianchi* 《蓮池大師全集》Volume 2. Shanghai: Shanghai Ancient Books Publishing House 上海古籍出版社.

Nāgārjuna. 龍樹. 《大智度論》 (*Great Treatise on the Perfection of Wisdom*) (*Mahā Prajñāpāramitā śāstra*), translated into Chinese by Kumarajiva, in 《日本大正新修大藏經》 (*Taisho Tripitaka*), Volume 25, page 155b. *Chinese Buddhist Electronic Text Association* (CBETA) http://www.cbeta.org/download/ebook.php (Accessed Nov. 28, 2020).

Wei, Dedong 魏德東著. 2016. "紐約的理性放生" ("The Wise Release of Life in New York City") in《中国民族报》 (*China Ethnic News*), Oct.25.

ENGLISH BIBLIOGRAPHY

Ho, Iris. 2012. "Turtle Release with Buddhist Blessings, Humane Education and Compassion: An alternative to Traditional Mercy Release." In *Humane Society International*, July 10. http://www.hsi.org/news/news/2012/07/buddhist_turtle_release_071012.html (Accessed Oct. 20, 2020).

Lombardi, Linda. 2014. "Building Good Karma: The Buddhist Ceremony of Releasing Turtles." In *Vet Street,* April 8. http://www.vetstreet.com/our-pet-experts/building-good-karma-the-buddhist-ceremony-of-releasing-turtles (Accessed Sept. 24, 2020).

Merchant, Carolyn. 1992. "Science and Worldviews." In *Radical Ecology*, pp. 41–59. New York: Routledge.

Nuwer, Rachel. 2012. "Buddhist Ceremonial Release of Captive Birds May Harm Wildlife." In *Scientific American,* August 1. https://www.scientificamerican.com/article/buddhist-ceremonial-release-captive-birds-may-harm-wildlife/ (Accessed Oct. 20, 2020).

———. 2014. "A Buddhist Ritual Gets an Ecologically Correct Update." Audubon, January–February. https://www.audubon.org/magazine/january-february-2014/a-buddhist-ritual-gets-ecologically (Accessed Oct. 20, 2020).

Westjan, Debra. 1997. "Buddhists Release Animals, Dismaying Wildlife Experts." November 1. https://www.nytimes.com/1997/01/11/nyregion/buddhists-release-animals-dismaying-wildlife-experts.html (Accessed Oct. 5, 2020).

IV

UTILITY OR SACRALITY?

TREES AND FORESTS IN CONTEMPORARY CHINA

Chapter 7

The "Ecological Forest of Daoism" in Minqin County, Gansu Province

Yang Der-Ruey[1]

Since awareness of human-caused ecological crises began to emerge in the West during the 1960s, Daoism has been widely acknowledged as one of the most eco-friendly religions in the world. Along with the deterioration of the global environment, there have been obvious failures on the part of incumbent political systems, both national and international, in dealing with our fast-approaching, global environmental disaster. Philosophers and historians of religion have striven to look deeper into non-Western religious traditions for alternative wisdom that might be helpful in saving the human species from extinction. The pathbreaking series of conferences on "Religions of the World and Ecology" organized by Mary Evelyn Tucker and John Grim at the Center for the Study of World Religions at Harvard University from 1996 to 1998 exemplifies this kind of endeavor.[2] One result of this conference series was the 2001 publication of the monumental anthology *"Daoism and Ecology: Ways within a Cosmic Landscape"* (Girardot, Miller and Liu 2001). This anthology and the high-quality works continuously offered by its contributors after 2001 have inspired a devoted intellectual enterprise in this field. Chinese historians and philosophers of Daoism in China have published at least nine books, fifteen journal articles, and four dissertations directly concerning the relationship between Daoism and ecology since 2004. Meanwhile, more than three hundred English-language articles and book chapters have been published in this field during the same period. Thanks to this burgeoning body of literature, the ecological implications of various aspects of Daoism, including its cosmology, ontology, epistemology, spatial/locative senses, ethics, and its esoteric learning of *fengshui* and inner alchemy, have been scrutinized and expounded. The different analyses in this burgeoning field are sometimes highly creatively or even fantastical, but most often favorable about Daoist culture in tone.[3]

In sharp contrast to the abundance of textual research and Daoist philosophical hermeneutics, empirical studies on the actual performance and practice of Daoist religious ethics on the ecological front are in short supply. A few journalistic articles, policy declarations, and official news releases by the mouthpieces of formal organizations such as the Chinese Daoist Association and the Alliance of Religions and Conservations address this issue. Yet it is very hard to find appraisals grounded on substantial empirical evidence about how Daoist practice today actually functions to advance a green agenda. The reforestation project facilitated by the Heilongdawang Temple in Shaanbei (northern Shaanxi Province), which was extensively studied first by anthropologist Adam Chau (2008), and later by sociologist Peng Shangqing (2015), is probably the only substantive example that can be used to testify to the green potential ascribed to Daoism by textual studies. Certainly, the conservation work being conducted by local Daoist organizations at the famous Daoist "auspicious caves and heavenly locales" (*dongtianfudi* 洞天福地), such as Maoshan in Jiangsu Province and Qingchengshan in Sichuan, and the encouraging stories of local temples are mentioned in several existing works. However, they are given little more than a scant introduction or a few passing notes. Therefore, it might be fair to say that, despite all the hopes and aspirations for Daoist environmental practice, the potential of Daoism to bring about a greener future for the globe has not yet been confirmed. In short, whether or not these interpretations are faithful to real-life Daoism as it existed in the past, and as it continues to exist in the current socio-political context, is a matter up for debate.

In light of the situation described earlier, this paper attempts to provide an empirical study of an eco-conservation project run by a Daoist organization. In line with the case of Heilongdawang Temple examined by Adam Chau (2008) and Peng Shangqing (2015), this case study is also situated in northwest China, and involved a temple and a reforestation project. However, unlike the undoubted success of the Heilongdawang Temple, as shown in chapter 8 of this volume, my case of the "Ecological Forest of Daoism" (*Daojiao shentailin* 道教生態林) project located in Minqin County, Gansu Province (甘肅省民勤縣), has been a failure. While most people prefer to focus on success stories, failures are often very instructive.

By presenting a failed case, this chapter does *not* intend to argue that Daoism is incapable of promoting ecologism. Rather, it aims to acknowledge that Daoism exists in a specific historical, local, and socio-political context. Indeed, it is inextricably intertwined with a unique ideational system consisting of doctrines, ideas, spiritual cultivation practices, and aesthetic styles, but it is not just an ideational system. First and foremost, Daoism is a more or less stable constellation of persons, activities, speeches, texts, symbols, resources, money, power, rules, artifacts, and tangible things. Whereas "ideational Daoism" can be embraced and embodied by an individual Dao-ist convert as her/his personal faith, real-life Daoism is more a sector of public social institutions than it is a private affair. Therefore, no matter how consistent ideational Daoism may be with what Arne Naess calls "deep ecol-ogy" (1989), which de-centralizes humans in the flow of natural forces, we will never know the potential of Daoism for advancing green agendas if we do not analyze its position and role in the here and now. Starting from this understanding as a starting point, this chapter argues that we would do better to hold realistic expectations about the potential of actual Daoism in con-temporary China in promoting a green agenda. As will be shown further, the vicissitudes of the "Ecological Forest of Daoism" project prove that Daoism, as a marginal social force in modern China, does not have the autonomy to influence broader ecological initiatives. Though Daoism has been ascribed some honorable duties by the state, little consideration has been given to the meager means at its command. Meanwhile, the religion itself is hardly allowed to pursue its own agenda. In fact, Daoism is literally deterred from serving its traditional public role because it is still regarded by many govern-ment officials as "superstitious," in the pejorative sense imported from the West. Daoist actions can easily be interpreted by Chinese officials and most contemporary secular urbanites as offending the alleged state/religion separa-tion principal of modernity. The viability of present-day Chinese Daoism to push forward a green agenda depends heavily on the state's will, as well as on environmental conditions. When ecological calamity is imminent or has already occurred, as with the case in Minqin, a politically castrated Daoism is often unable to help.

I. BACKGROUND: DESERTIFICATION OF THE AREA

Minqin County is a part of Wuwei City in Gansu Province in northwest China. Located in the northeastern tip of the Hexi Corridor (河西走廊), it covers an area of 15,900 square kilometers. The coordinates of the county are 101°49'41"–104°12'10"E and 38°3'45"–39°27'37"N. The proj-ect "Ecological Forest of Daoism" is situated on Suwu Hill (蘇武山), which is about twelve kilometers to the southeast of Minqin's county seat (figure 7.1).

From the map of Gansu, we can clearly see that Minqin County resembles a lonely tooth jutting out into the encroaching Gobi Desert of Inner Mongolia Autonomous Region to the north. The map nicely represents Minquin Coun-ty's role as a precarious remaining fortress of the more humid and agricultural eastern and southern part of northern China holding out against the invasion of the dry nomadic desert terrain of northern and western China.

民勤縣

Figure 7.1   Map of Minqin County (in red) in Gansu Province, holding out against the encroaching gobi desert of Inner Mongolia. Map by Oliver Teernstra.

Since the 1990s onward, Minqin has been widely acknowledged by Chinese people as a crucial practical and psychological garrison for guarding against the desertification descending from Mongolia. The Chinese government tends to project desertification as a military invasion progressing from the north—effectively appealing to an anxiety that has been deeply rooted in Han Chinese collective memory of nomadic pastoralist invasions for millennia. As a result, the war propaganda-esque slogan, "We can never let Minqin become the second Lop Nor! (絕不讓民勤成為第二個羅布泊)," authored by Premier Jiabao Wen in 2001, can still be seen painted on public walls and buildings in Minqin today.

However, if we seriously investigate the causes of desertification in the area, the previous depiction of the situation might be reversed. Many environmental historians argue that desertification in the Hexi Corridor, and especially in the drainage area of the Shiyang River, has been mainly caused by over-population and the over-development of agriculture underway since the early eighteenth century onward (Cheng, H. 2007; Zhang, X. et al. 2018; Wu, L. et al. 2020). Notably, there is evidence to suggest that rainfall in the surrounding regions has indeed been in decline over the past six centuries

(Yang, Bao. et al. 2011). So, it is actually Han Chinese people, with their thirsty agricultural mode of economy, who invaded this fragile oasis of Minqin County from the southeast, and finally destroyed it.

According to archival records, the population in the whole "Corridor to the West of the Yellow River," another name for the Gansu area, rarely exceeded 0.3 million from the second century BC until the late sixteenth century AD. However, the population size rapidly bourgeoned to 2.64 million in a survey conducted in 1776, and the same trend continued until it reached 3.18 million in 1851. By 1880, due to the massacre which occurred during and after the "Rebellion of Hui Muslims (回亂)" (1862–1873) and a series of small-scale domestic wars which took place during the early twentieth century, the population had dropped down to 0.8 million. It stayed below 1.2 million until the founding of the People's Republic of China in 1950. Since this time, however, the population has bourgeoned once again. In a 1978 survey, the population had reached a new historic height of 3.58 million. By 2004, it had reached 4.86 million. (Cheng 2007: 193–195, Table 6-37~44)

Historically, the Han Chinese were an agrarian civilization, practicing irrigated agriculture since ancient times. A huge population of Han Chinese would thus need a massive amount of water for irrigation. Likewise, the ratio of the quantity of utilized water to the quantity of total water resource, in this area rose abruptly from 17.4 percent in the mid-sixteenth century to 44.9 percent in 1776 and 46.9 percent in 1851 (Cheng 2007: 212–213, Table 6-55). The utilization rate of water resources corresponds to the historical change in the region's population size. In other words, the water needed to support the traditional agricultural economy had already exceeded the natural water supply's threshold for self-repair (35 percent according to UNCOD 1978) by the eighteenth century. Due to advances in irrigation technologies and the improvement of relevant infrastructure since the 1950s, the utilization rate had increased to an alarming 80 percent and above in 1966. The situation in the Shiyang River drainage area is undoubtedly the worst case in the whole region. The area's utilization rate had already reached 144.4 percent in 1950, but continued to rise to 188.1 percent in 1978 (Cheng ibid.) In 2004, this most important lifeline of water for Minqin County shockingly stopped flowing.

Symptoms of chronic drought in Minqin County in the early eighteenth century were recorded for the first time in the 1749 edition of the local gazetteer (Zhang 1749). Among the numerous hints contained in the gazetteer about the gradually deteriorating water supply, the most telling is the mention of three important conflicts between groups of farmers due to the shortage of irrigation water, occurring from 1722 to 1743, in 1725, and in 1727. Regrettably, these red flags were completely ignored. Baitinghai Lake (白亭海), downstream from the Shiyang River, was a large lake covering 4,000 square kilometers and reaching a depth of 60 meters in 1949. By only 1957 it had dried up. Subsequently, the Badain Jaran Desert to the northwest of Minqin and the Tengger Desert to the east of Minqin joined together to besiege the Minqin oasis. A few years later, the forest on nearby Suwu Hill disappeared and what was once a sanctuary for wildlife became a bald hill.

The Chinese government did not tackle the problem of desertification seriously until the United Nations organized a conference on desertification (UNCOD) in Nairobi, 1977. Two years later, the Chinese government launched a monumental project entitled "The Construction of Shelter-Forests in Northern, Northeastern, and Northwestern China" (三北防護林工程). This project marks the beginning of the Chinese government's endeavors to curb desertification. However, as could be expected, the scale of the problem and the speed of deterioration were much larger and faster than the scope of the project. It was not until 1996, during the first year of the Ninth Five-Year Plan (1996–2000), that the government began to address desertification in Minqin anew. According to the Ninth Five-Year Plan, Gansu Province would work to control desertification on 2,420 square kilometers of land, and refor-est 440 square kilometers. As a result of this political mission, the "Ecologi-cal Forest of Daoism" Project was born.

II. THE BIRTH OF THE "ECOLOGICAL
FOREST OF DAOISM"

For a rather poor county like Minqin, achieving the desertification control targets set by the higher authorities is a very challenging task, due to lack of personnel and funds. Thus, County leadership could only turn to mobilizing all potential resources in local civil society, including religious communities, to fulfill the goal. As a consequence, the leadership of the Minqin County government proposed a deal to influential local Daoist lineages. Simply put, the idea was that the County government would release the usage rights of the land surrounding Suwu Hill to the local Daoist community. This meant that they could rebuild Suwu Temple on the hill and establish it as the only legal site for conducting public Daoist religious activities in Minqin County. In exchange, the Daoist community would take responsibility for planting trees and grasses on the hill. The Minqin Daoist Association was hastily founded in response. Despite all the doubts and questions about its level of representa-tion, the Daoist Association acted as a legal body and signed a deal with the County government. The Minqin Daoist Association obtained fifteen square kilometers of desert and dunes from the government in 1996 and started to raise funds to fight desertification on the land.

How did the leadership of Minqin come up with such an idea in the first place? Why did they not try to cooperate with other religious organizations in other historically renowned sites? One possible explanation is that the nationwide Chinese Daoist Association had just announced *The Declaration of the Chinese Daoist Association on Global Ecology* in 1995, under the influence of the Alliance of Religions and Conservations (Lemche 2010: 21–22). Whether sincere or not, this official declaration demonstrated willingness on the part of Daoist organizations to respond to state environmental initiatives. Though the declaration is a meaningful background for the emergence of the project, to really explain why the project in Minqin answered the Chinese Daoist Association's call for ecological consciousness, we need to look into the history of Daoism and its situation in Minqin.

Minqin has been home to a consistent salient Daoist presence ever since its founding in the late fourteenth century. Although Minqin was occupied by Han Chinese people from 121 BCE to the third century CE, and then from 674 to 783 CE, the imprint of Han Chinese culture was almost completely eradicated by Turkish, Tibetan, Uyghur, and Mongolian settlers who took turns ruling the area from the eighteenth to the thirteenth century. Due to the cruel warfare between the Han Chinese and the Mongolian Empire during the late fourteenth century, Minqin had become a no-man's land by the end of the fourteenth century. Minqin was founded by the first emperor of the Ming Dynasty in 1386 as a military outpost tasked with guarding against Mongolian invaders. Zhenfanwei (鎮番衛), literally meaning "Outpost for Suppressing Barbarians," was the original name of Minqin, and this name lasted until 1928.[4] It was a community consisting almost exclusively of Han Chinese soldiers, military officers, and their families.[5] For such a community to survive in the dangerous borderland thousands of miles away from home (most of the officers came from the heartland of the Ming Dynasty–the Yangtze River Delta), spiritual support was just as important as material logistics. It is widely known that the royal family of the Ming Dynasty had a notable inclination toward Daoism. In fact, the royal lineage of Prince Su (肅王), who held the Emperor's mandate to control the entirety of Gansu from Lanzhou City, was descended from Zhu Ying (朱楧), the founder of the largest Daoist temple, the Temple of Golden Heaven (金天觀) in northwest China during the Ming Dynasty (Wu 2009: 396–399). Zhu Ying's allegiance to Daoism was inherited by his offspring, with eight out of the sixteen princes of this lineage taking Daoist names.[6] In order to boost the soldiers' morale and strengthen their sense of Han Chinese nationalism, which was crucial for consolidating their allegiance to the Ming Dynasty, a series of Han Chinese-style altars, shrines, and temples were built in the Zhenfanwei Castle and other government buildings. Apart from a few Confucian temples and altars or shrines for practicing the "state religion" of the Chinese empire, such as an altar for the God of Agriculture (*xian nongtan* 先農壇), most of the temples were dedicated to worshiping Daoist gods—especially the Daoist Gods of

War, such as the Emperor of Northern Heaven (玄天上帝), Emperor Guan (關帝), and the Ancestral Master of Thunder (雷祖).[7] Despite minor damage caused by warfare during the transitional period between the Ming and the Manchurian Qing Dynasty in the mid-seventeenth century and later during the Rebellion of Hui Muslims (1862–1873), the infrastructure of Daoism remained intact in Zhenfanwei for more than five centuries, until the establishment of the People's Republic in 1949.

Corresponding to its predominance in terms of temple buildings, Daoism also dominates seasonal festivals in Zhenfanwei. As a series of local gazetteers have shown, almost all popular festivals celebrated by the local population are derived from Daoist mythology,[8] such as the birthday of the Three Magistrates (三官) on the full-moon day of the first, seventh, and the tenth lunar months, the birthday of the Emperor of Northern Heaven on the third day of the third lunar month, the birthday of the City God (城隍) on the eighteenth day of the fifth lunar month, and the birthday of Emperor Guan and the Ancestral Master of Thunder on the twenty-fourth of the sixth lunar month.

Although Daoism, as an assemblage of Han Chinese culture, may have been a threat to Manchurian rulers, the Qing Dynasty did not try to suppress Daoism in Zhenfanwei. In fact, the number of Daoist temples in Zhenfan County increased during the Qing era. However, the situation began to deteriorate after the advent of the Republican Era (1911–1949), due to the continuous turmoil caused by intermittent Muslim riots and the modernist ideology of the ruling Nationalist Party. The Republican regime was fragile in northwestern China; it could do little harm to civil Daoism. Therefore, even though the scale and quality of Daoist temples and festivals regressed after the Qing era, priests and the common people could still practice Daoism without state harassment. That is, until the socialist revolutionary regime of 1949 changed almost everything. All the public expressions of Daoism, such as temples and festivals, were eradicated during and after the late 1950s.

In Minqin, some Daoist ritual practices, mainly Daoist funeral and memorial rites (for ancestors) resumed soon after Deng Xiaoping toppled the "Gang of Four" in 1976. It is still common custom in Minqin to invite a Daoist troupe to perform funeral rites for the deceased. According to my informants (who are Daoist priests themselves), there are at least nine renowned Daoist lineages in Minqin, including the Duans (段), the Huangs (黃), and the Zhangs (張). These lineages have recently trained about 80 or so ordained Daoist priests and more than 140 Daoist musicians in Minqin. These 200 or so Daoist personnel constitute several dozen Daoist troupes, which perform funerary, pacifying, healing, and blessing rituals for their clients. These troupes normally diagnose *fengshui*, choose auspicious dates, and give names to newborn babies upon clients' requests. As a result, most of the priests (and some good musicians too) have extremely busy schedules. It is quite common for a single individual or troupe to have more than twenty appointments per month. Therefore, the average monthly income of an ordained Daoist priest is 8,000 to 10,000 RMB, which is 2.5 to 3 times the standard salary of an ordinary wage-laborer in Minqin. Most younger Daoist priests invest in small businesses dealing in ritual supplies (such as paper money, paper crafts, incense, or candles), which are tasked to their wives. In Minqin, it is not unusual to meet a Daoist priest driving a Land Rover and owning a few decent flats in town (figures 7.2 and 7.3).

Despite the significant role Daoism has played in the history of Minqin and its continuous relevance in the common people's lives, not a single Daoist temple has been revived since the end of the Cultural Revolution (1966–1976). Given the relatively affluent position of Daoist priests and their clients, what the local Daoist community seemed to lack in order to revive its temples in Minqin was government approval. In this sense, it seems natural that the County's leadership would propose a deal to the Daoist community–granting authorization to revive a temple in exchange for efforts toward reforestation. Although it was a risky gamble, the local Daoist community still regarded this overture by the local government as a precious opportunity to finally get approval to build a temple.

Figure 7.2 Scene from a typical daoist funeral in Minqin County. Photo by Author.

Figure 7.3   Daoist priest presiding over funeral. Photo by the Author.

III. THE LIFE AND DEATH OF THE "ECOLOGICAL FOREST OF DAOISM"

The lifespan of the Ecological Forest of Daoism can be divided into three stages: (1) the end of 1996 until April 2003; (2) April 2003 until September 2008; and (3) September 2008 to the present day. During the first stage, the local Daoist community assumed the task of combating desertification in return for control of Suwu Hill. They raised around 200,000 RMB in funds (roughly 24,000 USD according to the exchange rate at the time) from local Daoist priests and their wealthy clients. The funds were first invested in infrastructure for the project, such as a well, 1 km of simple motorway, and 8 km of power lines. Other expenses included installing sand/gravel stabilizer on about 1 square kilometer of land, and purchasing and planting 31,000 tree seedlings (*Populus/Turanga* and *Elaeagnus angustifolia*) and 128,000 seedlings of bushy plants (*Sacsaoul, Caragana korshinskii, Hedysarum scoparium*). In addition, as they had explained to donors that some amount of their money would be used for the construction of a Daoist temple, they did spare funds to build a humble brick shrine on the hill, named Suwu Shrine. According to my informants, this shabby shrine was originally a storage unit for farming tools and a base for farm workers, which was later rebuilt into the current Daoist temple after additional funds were collected from local donors.

The political context in which the project was initiated changed following murky corruption allegations against the leadership of the local government in 1998. The whole project, and especially the make-shift Suwu Temple, was taken to be evidence of corruption. Luckily enough, the allegations and the resulting replacement of officials in the local government did not seriously sabotage the project. To the contrary, successors of the deposed officials tended to take a "benign neglect" policy toward the project in order to avoid any possible allegations of wrongdoing. Later in 2000, the central government organized a nationwide survey to assess the result of the anti-desertification mission of the Ninth Five-Year Program (1996 – 2000). The horrible situation of the Shiyang River drainage area shocked the then Prime Minister Wen Jiabao, leading him to author the motto, "We can never let Minqin become the second Lop Nor!" Under his command, the provincial government of Gansu pledged to make renewed efforts in its fight against desertification during the Tenth Five-Year Program (2001–2005). These events marked the second stage of the project's history.

From 2002 onward, the Wuwei City government and the Minqin County government reviewed the role of the Ecological Forest of Daoism in their collective measure toward desertification. As a result, they decided to try upgrading the Ecological Forest to a nationwide enterprise, making it into a model for state-society joint ventures in environmental protection. Thanks to the successful lobbying of local officials, the national-level Chinese Daoist Association was persuaded to join the project. In April 2003, the deputy chair and the secretary general of the Chinese Daoist Association, along with a few Chairs of Daoist Associations from other provinces such as Qinghai and Zhejiang, attended the opening ceremony of the "Base for the Construction of the Ecological Forest of Chinese Daoism" (中國道教生態林建設基地). Apart from the officials, hundreds of government employees, Daoist priests, and even common people were mobilized to participate in the event, which was staged at a remote corner at the edge of the desert. The preplanned event was reported widely by all sorts of news media.

Accompanying the project's fresh start was a new wave of fund-raising campaigns that swept across the whole of China. Now headed by a national-level Daoist Association instead of a tiny local Daoist organization, the project raised more than 3.3 million RMB (roughly 437,000 USD according to the exchange rate in September 2007) during the next four and a half years. This figure is 16.5 times (or 18.2 times in USD) the total donation the project obtained during its first stage. Thanks to the budget increase, the project infrastructure's roads, electrical supply, and its network of underground water pipes, were greatly improved. The project enlarged its reforestation area to cover 6.5 square kilometers of land and planted more than 500 thousand seedlings of trees and bushy plants. A spatial planning team was hired to design an overall plan for the entire fifteen square kilometers of land. The design team produced an ultra-utopian plan that included not just the Suwu Temple, a forest, and bushland, but also two ponds, one cemetery, a hotel, a library, an old-age activity center, a juvenile activity center, and even a camel racing course (figure 7.4)!

Regrettably, this dream-plan was shattered by the drying of the Shiyang River, which occurred just a few months after the plan was produced. The river disaster led the government to issue its strongest ever ban on extracting underground water. Accordingly, all existing wells in Minqin were to be blocked and pumping machinery confiscated within a few months' time. Every household and every corporation in Minqin, including the Ecological Forest of Daoism, would have to pay for water exclusively supplied by waterworks. The prohibitive costs led the overall plan of the Ecological Forest to quietly shrink at the end of 2004. Plans for the Suwu Temple, forest, bushland, and cemetery were preserved, but the hotel, library, Senior Center, Youth Center, camel racing course, and especially the two ponds were all dropped.

In March 2005, Master Ren Farong (任法融), the chairman of the Chinese Daoist Association, attended the inauguration ceremony for the construction of Suwu Temple. Three and a half years later in September 2008, the Suwu Temple was completed. This "revived" Suwu Temple is a temple for all the key Daoist gods who have ever been widely worshiped by the local people.

Figure 7.4 **Plan for Suwu hill's development.** Photo by Author.

Its most important buildings are for honoring the Emperor of Northern Heaven and Emperor Guan. Other Daoist gods who have long histories in Minqin County, such as the Ancestral Master of Thunder, the Dragon King, the Medicine King, and the City God, were all given seats beside the two Emperors. The tiny Suwu Shrine was relegated to the eastern flank – the lowest and least decorated building of the temple complex (figure 7.5).

The inauguration ceremony for Suwu Temple held in September 2008 marked the end of the second stage of the Ecological Forest of Daoism. In sharp contrast to the dramatic movement from dormancy to expansion witnessed at the end of the first stage of its existence, the second stage ended in a stealthy contraction of project plans. Suwu's final stage would be characterized by a silent and slow death caused by the drying up of resources—both water and money.

The well that the project relied upon was banned by the government in 2006, meaning that the project had to pay several thousand RMB each month to buy water. For a nonprofit organization like Suwu Temple, this situation was simply unsustainable. Although it is true that Suwu Temple did manage to attract some amount of donations from local Daoist priests and pious believers through organizing an annual temple fair, this income is barely enough for the maintenance of the temple itself. Consequently, the funds

Figure 7.5 **Bird's-eye view of Suwu Temple.** Drone photo by Tao Jin.

raised from 2003 to 2007 gradually ran out without any further replenishment. Sometime around 2014 or 2015, the project stopped paying wages to farm and construction workers and they left for good. Now, the Daoist Association of Minqin, which is located inside Suwu Temple, is still trying their best to water the trees beside the temple and the road leading to the temple with the meager budget that they still have. As for the forest and bushland cultivated during the preceding few years, the Daoist Association of Minqin can do nothing but pray for the mercy of the Dragon King to bring rain. Although the project itself was not a success, after twelve years (1996–2008) of laborious and costly collaboration with the government, the local Daoist community got what it wanted at the start of the project—government authorization to build a Daoist temple (figure 7.6).

Why did the local government stop pushing the Daoist Association to invest in the continuation of the Ecological Forest? One possible reason is that the government has greatly improved their knowledge about the causes of desertification and vastly changed their approach toward problem. Instead of trying to mobilize the population to tackle a desertification with manual labor and determination (an approach which the Chinese Communist Party was terribly good at in the past, as shown in Adam Yuet Chau's chapter 8), the twenty-first-century Chinese government is increasingly approaching its

Figure 7.6   **The current environmental situation at Suwu Hill.** Drone photo by Tao Jin.

environmental objectives with new science, technology, and economic incentives. In terms of the desertification of the Shiyang River drainage area, the Chinese government has been mobilizing earth scientists with large national budgets to study the area's desertification problem since 2000. Their research has gradually, but fundamentally, changed the government's perspective on the problem and its strategies. In summary, research in the Shiyang River drainage area led the government to acknowledge that over-population and over-development of traditional agriculture were the central causes of desertification. This understanding has helped the government understand that it must tackle these two causes resolutely and immediately. In response, the government began to (1) relocate "ecological migrants" in order to reduce the population living in this area; (2) force existing family farms to adopt advanced water-saving farming technologies; and (3) induce farmers to grow water-saving, high market-value crops, such as drought-resistant medical herbs, dates, and wine grapes. This new approach was first integrated and authorized into official policy as a part of the Eleventh Five-Year Plan (2006–2010). It was then further systematized through the Program for Tackling the Key Issues of Shiyang River Drainage Area, which was authorized at the end of 2007. This program, with a half-billion RMB budget, not only attempts to regulate the region's population and agriculture, but also plans to conduct massive-scale infrastructure construction involving several rivers. As a result, at least in terms of desertification control, there is no more role left for the local level governments, like those of Wuwei City or Minqin County, to propose any interventions. The only responsibility left to local officials and civil society is the efficient execution of commands descending from the central or provincial government. What concerns the local government of Minqin now is no longer the Ecological Forest of Daoism, but the new vineyard for growing wine grapes that lays beside it.

IV. DAOIST RESPONSES TO CALAMITY

It can easily be ascertained that the Daoist community in Minqin feels disappointed, frustrated, and even angry about the ever-developing desertification of the area surrounding the Suwu Temple. After all, many of them were involved in the reforestation project and contributed something to it over the past two decades. However, just like most ordinary people, they would rather avoid looking at the appalling reality in favor of trying to keep their lives going as they always have. Hence, it is to be expected that the Ecological Forest of Daoism has become a topic of discussion that is avoided by local Daoist priests. Their typical response to the issue is, firstly, to apologize for their ignorance or bad memory about it, and then try to steer the conversation toward things pertaining to the Suwu Temple. Despite all the difficulties the temple has faced since it was built—from attracting worshippers to preserving the surrounding plantation—local Daoist priests still consider the temple a proud accomplishment of Minqin's Daoism. Nevertheless, their deeply repressed frustration and anger about the ongoing environmental calamity or, more accurately, the stupidity and impotence of local leadership in handling the calamity is inescapably revealed through their pessimistic forecasts about the future of the temple. Consensus has it that, unless some miracle happens, the Suwu Temple will be devoured by sand in a few years' time.

Will there be a miracle? Like most people in China today, Daoist priests in Minqin do not show any unusual propensity for imagining miracles. At least, I have never heard a Daoist priest mention an unrealistic, fantastic, or supernatural method that might bring water to Minqin and stop the invasion of desert. Indeed, Daoists do at times have a propensity to make moral comments decrying official corruption, avarice, and excess as additional factors contributing to the state of the environment. However, this inclination can hardly be seen as something stemming uniquely from their religion since it reflects the usual rhetorical style widely used by ordinary people in Minqin. What is unique about Daoist priests in Minqin is that they do *not* try to propose religious solutions, such as ritual penitence or prayer for forgiveness, to mitigate the disaster. If this is really the way they exemplify the Daoist teachings of "wordlessness" (*bu yan* 不言) and "nonaction" (*wu wei* 無為), then we cannot help but acknowledge that they have all achieved a very high level of spiritual enlightenment. However, a copy of the *Scroll of Dragon Kings* (*Longwangjing* 龍王經)[9] in the library of an aged Daoist master clearly indicates that the "wordlessness" and "nonaction" of Daoist priests in today's Minqin may have a more persuasive explanation not found in ancient classics such as the *Dao De Jing* or the *Zhuangzi*. Why? The scroll is the core text of the Daoist rainmaking ritual, that is, the method for initiating a miracle from the gods.

According to Master Duan, the most recent Daoist master in the Duan lineage—a renowned Daoist family in Minqin for generations—the *Scroll of Dragon Kings* is one of the texts he hand-copied from his father's collection in the late 1970s or early 1980s. Despite the generally poor conditions for maintaining his collection of ritual texts, the condition of the scroll is comparatively good because it has never been used in actual ritual occasion since it was made. The last time Master Duan witnessed a rain praying ritual was in the early 1960s, when he served as an assistant in the liturgy officiated by his father. As a teenage novice apprenticed with his strict father at that time, he could do nothing but concentrate on memorizing the liturgy itself. Consequently, he is confident that he can still perform the entire liturgy accurately although he has not had a single chance to perform it himself or to witness a similar ritual during the past half century. However, the price of his good memory about the intricacies of the ritual corpus is that he can barely recount the social historical context of the event he witnessed in his youth. However, he does remember that there was a serious drought going on, and the rain did come soon after the ritual.

As the adverse political environment for religious practice after 1949 have already been explained earlier, there is no need to repeat the historical factors that made Daoist rainmaking rituals disappear from Minqin after the 1960s. Actually, if it were not for the horrendous drought and famine that befell Minqin during the nationwide "Three-Year Natural Disaster (三年自然災害)" (1959–1961), which weakened the state's control over local religious activities,[10] it would have been virtually impossible for Master Duan to participate in a rainmaking ritual officiated by his father. According to Master Wu, an heir of another locally renowned Daoist family who is now around fifty years old, the last properly observed rainmaking ritual in Minqin was actually much earlier than the one officiated by Master Duan's father. Master Wu claims that a large-scale rainmaking ritual initiated by the mayor of Minqin and officiated by his grandfather during the Republican Era (1912–1949) should be regarded as the most recent authentic Daoist rain ritual.[11] If Master Wu is correct, then a Daoist rainmaking ritual has not been properly performed in Minqin for more than eighty years.

In a place that has been tortured for so long by drought and famine, how is it that the rainmaking ritual of the mainstream religion of the local community has been suspended for eight or even five decades? In fact, if it were not for a frustrated and bored anthropologist who happened to discover the *Scroll of Dragon Kings* in a retired Daoist master's ritual manual collection on a banal afternoon, and started to ask questions about it, all knowledge about the liturgy may well be buried in some dark corner of the aged master's memory forever. The oddity of this situation is even more salient if we take into account the fact that most of the other rituals in local Daoist liturgical tradition, especially death rituals and earth-pacifying rites, were never completely impeded, except during the ten years of the Cultural Revolution. In fact, the rainmaking ritual, as a classical means to plead for divine intervention and the rescue of the local community from natural disaster is one of the most politically sensitive rituals in the Daoist liturgical tradition. To clarify this point, we need to take a closer look at the structural features of the ritual.

The entire procedure of the ritual unfolds in the following way. First, the initiator or sponsor formally invites the Daoist master to officiate a ritual by following classical etiquette. After the whole process is set out and while preparing the needed paraphernalia and offerings for sacrifice, the team of Daoist priests, the initiator (and his/her delegates), and, to a lesser extent the entire community, must fast or observe vegetarianism. Animal slaughter is strictly banned in the community in preparation for the ceremony. On the ordained day of the ritual, the Daoist master starts the liturgy with a literary oral report to the Daoist pantheon of gods who have been invited down to attend the ritual, about how terrible the drought is, how the leaders and commoners of this community sincerely regret their wrongdoings, and how they desperately beg for the gods' pardon and salvation. Then, the Daoist troupe sets out to fetch water from the key water source of the community, such as a spring, a pond, a well, or a river. The water will then be poured into four vases on an altar. Having done this, the Daoist troupe then begins to recite the *Scroll of Dragon Kings* several times while the ceremonial master sanctifies the water in vases. Next, the master uses a willow branch to sprinkle the water around the ritual site while the troupe chants the *Mantra of Mulang* (木郎咒).[12] After that, the Daoist troupe returns to the water source again and drops an intricate Daoist iron talisman into the water. This action is followed by a period of anxious waiting time for rainfall. The waiting period may take several hours or a whole day. If rain does come, then the whole liturgy will end happily with a thanksgiving sacrifice and the community will stop fasting. However, if the rain does not come, the frustration of the public might easily lead to the discontent targeted on local leaders, as people might start accusing certain people of influence of not being sincere in their penitence, thus offending the gods. "To be honest, I don't really know what to do next if no rain comes. I guess I would just do the liturgy again with even greater measures of caution," said Master Duan.

Based on the procedures summarized earlier, we can determine four key features that highlight the political sensitivity of the Daoist rainmaking ritual. Firstly, the ritual requires the local political leader(s) to act as the (nominal) initiator. Secondly, it requires both the political leader(s) and the common people to confess, repent, and, sometimes, punish themselves. Thirdly, it will unavoidably become a ritual related not just to the political leader's legitimacy, but also to the head Daoist master's charisma and religious authority. Finally, such a ritual would formally acknowledge the role of Daoism as a key sector in the political system and the Daoist clergy as primary political agents. This acknowledgment could prove very dangerous for Daoism and local Daoists if the established temporal authorities were to consider the ritual a transgression and a threat to their political legitimacy.

Daoist rainmaking rituals are meant to be initiated by the secular political leader of the region ravaged by drought. Theoretically speaking, neither an individual Daoist master, the clergy of a temple, nor any civilian, no matter how influential he or she might be in the local community, could have the right to act as the nominal "initiator of the ritual" (*zhaizhu* 齋主). Individuals might try to wield their influence to push a political leader to initiate the ritual, but it would be improper for anyone to begin the ritual without, at the very least, getting the endorsement of a political leader in advance. Actually, the rainmaking ritual is just one example of a huge portion of Daoist liturgy, which can be called "public rituals." This category consists of the regular rituals for paying homage to the massive pantheon who control the "Three Realms" (Heaven, Earth, and Water), rituals for pacifying and stabilizing the Earth, appeasing and sending off resentful ghosts who died of war or natural disasters, fending off natural disasters such as excessive rain, flood, hail, wild fire, and locusts, and the *Jiao* Ceremony (醮) for divine blessings for the whole community. These rituals are "public" because they are not aimed at protecting and improving the well-being of an individual or a family, but a whole community at large. Whether it be a village, a nation, or the entirety of humankind, Daoist public rituals act to maintain order and resolve any problems that may threaten it. According to the logic of Daoism, since the leader of the community is the public authority who holds the duty as well as the right to file a report to the Celestial Court of powerful deities about a given territory, it is his or her duty to initiate these public rituals. On such occasions, Daoist priests are both mediators between human magistrates and the Celestial Court earlier, as well as celestial messengers who mobilize magical means warranted by the gods to tackle a problem.

The principle of temporal authority as ritual initiator or sponsor, and Daoist clergy as mediators between temporal authority and the gods thus presumes a symbiotic, or collaborative relationship between a given political system and Daoism. Therefore, this relationship stands to be nullified whenever a political system may refuse to cooperate with Daoism. Ever since Buddhism began to rival Daoism in the fifth century this has not been a rare situation in Chinese history. As a result, we can witness innumerable cases where the local political atmosphere did not favor Daoism and Daoist public rituals were not initiated by local political leaders, but by influential civilians, corporations, or the Daoist clergy themselves. However, it should also be acknowledged that the principle could be allowed to lie dormant in peaceful times and preserved for occasions where the value or efficacy of the ritual being performed are deemed high, and the political risks are low. For example, it is very hard to tell whether the Ritual of the Yellow Register (*Huangluzhai* 黃籙齋) for appeasing and sending off wandering ghosts being held during the Ghost Festival (the full-moon day of the seventh lunar month) of a peaceful year is efficacious or not because there is no decisive criteria for judging its success. In sharp contrast, rainmaking rituals and all the other public rituals for fending off natural disasters are highly risky endeavors as their efficacy will be revealed immediately, and both political and religious legitimacy and authority will be at stake. All the community members, and especially Daoist priests, will be obliged to bet their reputations on the ritual. For this reason, Daoist priests tend to abide by the ancient precept requiring official initiation to conduct public rituals. This is probably the key reason why Master Wu claims that the rainmaking ritual officiated by his grandfather during the Republican Era is the last authentic one in Minqin. It was, in fact, most likely the last one that was *properly initiated* by the mayor of Minqin.

Apart from the high level of reputational risk involved, there is another innate reason for the need to have a political leader as the nominal initiator of a Daoist rainmaking ritual. According to Daoist theory espoused in ritual observances and canonical prayers, drought and all kinds of natural disasters are mainly caused by the wrongdoings and moral corruption of the affected community. In this context, the community's political leader is the one who should take the foremost responsibility. Therefore, aside from the virtue and the magical prowess of the leading Daoist master, the efficacy of the ritual also depends heavily on the sincere self-punishment, and repentance of a given area's political leader(s). Other community members are also encouraged to fast, to expose their nude bodies to the sun, abstain from sexual relations, and subject themselves in creative ways to pain and damage before and during the ritual, but they are not expected to do so as much as the leaders. Therefore, to push the incumbent political leader to initiate a rainmaking ritual amounts to requiring him or her to confess moral and political failure to the gods via the mediation of Daoist priests. Consequently, his sincerity and, more precisely, his willingness to sacrifice himself at the cost of his pride, health, or even his very life, in order to rescue his subjects, will be put to the test. The legitimacy of his rule may well be challenged if the rain does not come. Certainly, were such an embarrassing situation to happen, the political leader would likely find one way or another to relegate all the responsibility to the officiating Daoist master. However, scapegoating someone else is not an effective method for repairing damaged legitimacy. Even in holding the Daoist master accountable for the failed ritual, the political leader would have to pay a substantial hidden political cost, which in some dire situations could literally cost his or her life.

The aforementioned analysis clearly indicates that the rainmaking ritual, as one genre of public rituals for fending off natural and other disasters, is a classical occasion wherein the political nature of Daoism will be highlighted and the Daoist master will be elevated to the status of a symbolic leader of the local community. Although the transfiguration of Daoism and Daoist clergy on these kinds of ritual occasions is only a temporary, symbolic arrangement, it would nevertheless be viewed by the temporal political authority as threatening and suspicious. However, since Daoism can at least serve as a mechanism for boosting the morale of the people to fight for survival amidst calamities, successive Chinese dynasties developed an intricate ritual-symbolic system to incorporate and encompass Daoist crisis-fending rituals during the past two millennia. In contrast, in their insistent secularism or, radical Marxist-Leninist atheism, the modern Chinese state since the early twentieth century onward has increasingly demolished the traditional framework of symbolic encompassment that incorporated Daoist rituals of crisis. In other words, whereas in imperial China, political authorities were sometimes willing to take political risks to keep open the possibility of a miracle brought about by Daoist ritual, modern Chinese officials would never consider such actions.

V. CONCLUDING REMARKS

Viewed from a historical perspective, the vicissitudes of the Ecological Forest of Daoism in Minqin can be seen as the epitome of the embarrassing and disheartening situation in which Daoism finds itself in modern China. The traditional liturgies that Daoism once used to harmonize the nature-human relationship and to fend off ecological calamities cannot be enacted again because they all presuppose an encompassing symbolic/ritual order coalescing political authority with ritual or religious efficacy. These rituals can be sustained only by an imperial rule with a "Heavenly Mandate." By switching the source of its sovereignty from "Heavenly Mandate" to "the people," the modern Chinese state deprived Daoism of the possibility to act, even symbolically, as a parallel collaborator with the state in maintaining worldly order, through staging public rituals. Consequently, having been stripped of most of the assets, it had inherited as an accomplice to imperial rule, modern Daoism has been neutralized, forced to retreat to the private sphere for almost a century. The Daoist community in Minqin County has long embraced an entrenched apathy toward all sorts of public issues. Thus, the initiation of the Ecological Forest of Daoism can well be seen as an attempt to invent a new public role for Daoism that is compatible with the modern state. For this particular project, Daoism was expected to assume the role of modern NGOs. Ideally, it would have functioned to mobilize and organize a civil movement after the model set forth by charity initiatives and some public awareness campaigns. Unfortunately, despite serious efforts on the part of the Daoist community in Minqin, the depletion of the water supply led the project to an early demise. The frustrated Daoist community in Minqin left behind the failed project, and withdrew to the social position it was originally assigned by the modern state—once again practicing their religion exclusively in the private sphere.

Theoretically, it is possible that the Daoist community in Minqin turn itself into a modern nongovernmental organization for promoting ecological awareness among the citizenry. In other words, Daoism in Minqin might still be able to invent some kind of public role based on the credentials it has obtained via the Ecological Forest project. However, the reality is that Daoism has been gradually excluded from the public sphere once again as the project became increasingly neglected by the local government after 2008 with the inauguration of the central government's Program for Tackling the Key Issues of Shiyang River Drainage Area. An awkward event in May 2018 provides a particularly clear illustration of the Minqin Daoist community's silent withdrawal from the public sphere.

On May 2, 2018, the China News Service published a news article entitled "Daoist priests Practice Exorcism at the Inauguration Ceremony of the Thorium Molten-Salt Reactor Nuclear Power System Project of the Chinese Academy of Sciences." With a few photos and a short video clip of the scene, this article relates how a group of Daoist priests were called to perform a sacrificial rite on April 26, 2018, for exorcizing evil forces that may have been haunting the construction site of the thorium molten-salt reactor nuclear power system in Minqin. The author then harshly criticized the Shanghai Institute of Applied Science, the responsible unit of this project, for allowing local contractors to organize such a "superstitious" activity for such an advanced scientific project. This breaking news successfully attracted a large audience from all over China and stirred up a sizable wave of discussion on the internet. Interestingly, many netizens opposed the critical stance of the article, and expressed support for the freedom of local contractors to initiate the construction work using folk traditions. However, all the officials involved in the ritual, either as organizers or attendees, were punished as offenders of party discipline and government regulations. The Daoist priests who performed the ritual were exempted from punishment because there is no law forbidding them to provide service for governmental projects. However, it is not at all difficult to imagine why political elites in Minqin and the surrounding area might shy away from future interaction with Daoism for the sake of their careers.

NOTES

1. Nate Sims and Mayfair Yang contributed to the editing and polishing of this chapter.
2. https://fore.yale.edu/Event-Listings/Religions-World-and-Ecology-Conference-Series/Religions-World-and-Ecology-Archive (accessed Dec. 11, 2020).
3. See, for example, Miller, James (2017).
4. Zhenfanwei 鎮番衛 was renamed "Zhenfan County 鎮番縣" by the Manchurian Qing government in 1725.
5. Most of the soldiers claimed to come from Nanjing City, Jiangsu Province, and the "Big Huai Tree (i.e. Japanese Pagoda Tree)", a mythical place in Shanxi Province. See Wu (2009: 120).

6. For similar practices among royal family members during the Ming Dynasty, see Goossaert (2008), Wang (2012) and Zhou and Lu (2015).

7. For reference, in contrast to the six Buddhist temples/towers, seven altars/ shrines for the "state religion", and six temples/shrines of Confucianism, there are thirty Daoist temples recorded by a 1749 edition of the local gazetteer. The cast of temples and shrines recorded appeared again and again in the 1825, 1908, and 1934 editions. Daoist temples recorded in later editions always tend to outnumber the count of the previous one.

8. The "Buddha's birthday" on the eighth day of the fourth lunar month is the only exception.

9. The Zhengtong Daoist Canon (正統道藏) contains two different versions of Longwangjing (龍王經), one is the Taishang Dongyuan shuo qingyu Longwangjing (太上洞淵說請雨龍王經) in the Dongxuan Section (洞玄部) and the other is the Taishang Yuanshitianzun shuo dayu Longwangjing (太上元始天尊說大雨龍王經) in the Dongzhen Section (洞真部). The copy I saw in Minqin is rather similar to the latter version. The two versions are almost identical in terms of their structure. They both are a short story in which the highest god of Daoism, the King of Dao (*Daojun* 道君) or the Primeval Lord of Heaven (*Yuanshi Tianzun* 元始天尊), reveals to the oracle how to pray for rain. The oracle of the two texts is identical. Both texts consist of the names of Dragon Kings who control rainfall over the world, although the first version has fifty-eight names, whereas the latter has sixty-eight. The supporting casts of the two versions are completely different.

10. For a rather brief but broad report of the religious surge which occurred during the "three-year natural disaster" era (1959–1961), see *Zhongguo huidaomen shiliao jicheng bianzuan weiyuanhui* (中國會道們史料集成編纂委員會) (2004). Ann Anagnost (1987; 1994) also offered an anthropological account of a few cases which happened during that time.

11. Master Wu cannot give the exact year his grandfather officiated the large-scaled rainmaking ritual. Neither are we able to infer the possible time based on the time of serious drought in Gansu during the Republican era because this region was continuously ravaged by drought throughout the period.

12. Apart from the last sentence that only indicates its function as a mantra for commanding spirits, the Mantra of Mulang (木郎咒) reads like a long poem with astonishing literary grace. Allegedly being authored by Yuchan Bai (白玉蟾), one of the greatest Daoist master-cum-poet during the twelfth and thirteenth centuries, this mantra describes a series of spectacles in which a multitude of gods and heavenly messengers collaborate with each other in extinguishing all the evil forces causing drought and thus bring about a breathtaking thunder storm.

ENGLISH AND WESTERN LANGUAGES

Anagnost, Ann S. 1987. "Politics and Magic in Contemporary China." *Modern China*, vol. 13, no. 1, pp. 41–61.

Anagnost, Ann S. 1994. "The Politics of Ritual Displacement." In Keyes, C. F., Kendall, L. and Hardacre, H. eds., *Asian Visions of Authority: Religion and the*

Modern States of East and Southeast Asia, pp. 221–254. Honolulu: University of Hawaii Press.

Chau, Adam Y. 2008. *Miraculous Response: Doing Popular Religion in Contemporary China*. Stanford University Press.

Girardot, Norman J., James Miller and Liu, Xun, eds. 2001. *Daoism and Ecology: Ways Within a Cosmic Landscape*. Cambridge: Harvard University Press.

Goossaert, Vincent 2008. "Mapping Charisma among Chinese Religious Specialists." *Nova Religio*, vol. 12, no. 2, pp. 14–18.

Lemche, Jennifer. 2010. The Greening of Chinese Daoism: Modernity, Bureaucracy and Ecology in Contemporary Chinese Religion, M.A. Thesis. Canada: Queen's University.

Miller, James. 2017. *China's Green Religion: Daoism and the Quest for a Sustainable Future*. New York: Columbia University Press.

Naess, Arne. 1989. *Ecology, Community, and Lifestyle*. David Rothenerg, editor and trans. Cambridge, UK: Cambridge University Press.

United Nations Conference on Desertification (UNCOD). 1978. *Round-up, Plan of Action and Resolutions*. New York: United Nations.

Wang, Richard G. 2012. *The Ming Prince and Daoism: Institutional Patronage of an Elite*. Oxford University Press.

Wu, Lei, Changbin Li, Xuhong Xie, Zhibin He, Wanrui Wang, Yuan Zhang, Jianmei Wei, and Jianan Lv. 2020. "The Impact of Increasing Land Productivity on Groundwater Dynamics: A Case Study of an Oasis Located at the Edge of the Gobi Desert." *Carbon Balance Manage*, vol. 15, no. 7, https://doi.org/10.1186/s13021-020-00142-7.

Yang, Bao, Chun Qin, Achim Brauning, Iris Burchardt, and Jingjing Liu. 2011. "Rainfall history for the Hexi Corridor in the Arid Northwest China during the past 620 Years Derived from Tree Rings." *International Journal of Climatology*, vol. 31, pp. 1166–1176. Published online 14 April 2010 in Wiley Online Library (wileyonlinelibrary.com) DOI: 10.1002/joc.2143.

Zhang, Xunhe, Nai'ang Wang, Zunyi Xie, Xuanlong Ma and Alfredo Huete. 2018. Water Loss Due to Increasing Planted Vegetation over the Badain Jaran Desert, China. *Remote Sensing*, vol. 10, p. 134; doi:10.3390/rs10010134.

CHINESE LANGUAGE

Anonymous. 2004. *Taishang Yuanshitianzun shuo dayu Longwangjing* 太上元始天尊說大雨龍王經, in Zhonghua Daoist Canon 中華道藏, Vol. 6, p. 218, Beijing: Huaxia Chubanshe.

Anonymous. 2004. *Taishang Dongyuan shuo qingyu Longwangjing* 太上洞淵說請雨龍王經, in Zhonghua Daoist Canon 中華道藏, Vol. 30, p. 124, Beijing: Huaxia Chubanshe.

Cheng, Hongyi 程弘毅. 2007. *The Desertification of the Hexi Area in Historical Times* 河西地区历史时期沙漠化研究, PhD Thesis, Lanzhou University.

Peng, Shangqing 彭尚青. 2015. *"Belief Relations" and the Formation of Folk Elite: Based on the Belief of Black Dragon King in Northern Shannxi* ('信仰關係'與民間權威的構成: 以陝北黑龍廟民間信仰爲例). PhD Thesis, Eastern China Normal University.

Wu, Mu 武沐. 2009. Gansutongshi – Ming Qing Juan甘肅通史·明清卷, Gansu People's Press.

Zhang, Zhizhai 張治齋. 1749 (乾隆14年). ed. *Zhenfanxianzhi* 鎮番縣誌, in Wuliangkaozhi Liudeji Quanzhi, Vol. Ren (五涼考治六德集全志·仁集), Zhongguo Fangzhicongshu Vol. 3160 - 3161 (中國方志叢書, 冊3160-3161）, Taipei: Chengwen Publishing Company.

Zhongguo huidaomen shiliao jicheng bianzuan weiyuanhui 中國會道門史料集成編纂委員會. 2004. *Zhongguo huidaomen shiliao jicheng: jinbainianlai huidaomen de zuzhi yu fenbu* (中國會道門史料集成：近百年來會道門的組織與分布), Beijing: China Social Sciences Press.

Zhou, Leijie 周雷傑 and Lu, Min 路旻. 2015. "Mingchao Suwang xi Daohao Kaobian 明朝肅王系道號考辨", *Zhongguo Daojiao* 中國道教, no. 6, pp. 41–44.

Chapter 8

Homo Arborealus

The Intermeshing of Regimes of Tree-Mindedness

Adam Yuet Chau

I. INTRODUCTION: VARIETIES OF TREE-MINDEDNESS

Trees belong to the plant kingdom in our general division of the world of living things (the other being the animal kingdom). Trees are just trees, masses of organic matter with no intrinsic meaning. But of course, in the long bio-cultural evolutionary process of humans, we have developed an elaborate symbolic universe in which tress and their various components or aggregates (e.g. fruits, timber, leaves, woods, and forests) play an important part. These symbolic constructs vary greatly across cultures, time periods, and ecological zones (see Rival 1998). In the twentieth and twenty-first centuries, a number of distinctive "regimes of tree-mindedness" have emerged. By "tree-mindedness," I am referring to a kind of obsession about trees.[1] This obsession is collective, and is a product of systematic, long-term cultural inculcation or large-scale sociopolitical mobilization. The word "regime" in the expression is highlighting the normative nature of this obsession. Each regime of tree-mindedness evolves over time, and different regimes of tree-mindedness encounter one another, borrow elements from one another, compete and intermesh with one another. Lest the expression "mindedness" might inadvertently evoke an emphasis on conceptions and thoughts only, I would like to stress that every "regime of tree-mindedness" includes a constellation of practices that contribute to the maintenance and reproduction of the regime (e.g., planting trees; harvesting, transporting, and marketing timber; developing sylvan-botany and forestry science; "hunting" tree specimens[2]; cultivating bonsai; chaining oneself to trees to prevent loggers from cutting down trees; growing mushrooms on fallen and rotten tree trunks; building tree houses; and maintaining *fengshui* groves).

The growing global awareness of anthropogenic climate change has made us become tree-minded more than ever before, and trees are taking on increasingly sacred characters. Environmentalists have been leading campaigns to protect the world's remaining pristine forest covers, especially tropical rain forests, the latter seen as the last line of defense against the alarmingly expanding global carbon deficit. Even though some scholars have warned against problematic romanticism (e.g., Slater 2003), rain forests have continued to be held up as sacred objects or icons. The outrage caused by the largely deliberate mass burning of the Amazonian rain forests in Brazil in 2019 (for clearing the forests for farming, livestock raising, logging, mining, etc.) attested to the symbolic weight of rain forests in the global consciousness.

In China, environmentalism as we know it today first sprouted in the 1990s, when people began to realize the extent of the environmental damage wrecked by rapid and aggressive industrialization of both the Maoist and reform eras. Intellectuals and academics began forming nongovernmental organizations (NGOs) in pursuit of environmentalist goals, often based on Western models (e.g., the Beijing-based environmentalist NGO Friends of Nature was founded in 1994). These NGOs aim at educating the general public about environmentalism, working with the state in formulating relevant policies (e.g., environmental protection; conservation; renewable energy; etc.), and networking with international environmentalist organizations to push for broader-scale changes. Many of these environmentalist NGOs in China organize tree-planting activities on a regular basis. On the surface, these tree-planting activities seem to be an expression of a global-environmentalist regime of tree-mindedness; yet in reality, they inherit many important features from mass tree-planting practices from the high-socialist era (i.e., 1950s to 1970s), during which a different regime of tree-mindedness reigned.

While global environmentalism is primarily based on an increasingly sophisticated scientific understanding of habitats, the ecological system, and the atmosphere, we have also become increasingly more aware of the fact that in many cultures around the world there are very diverse, indigenous understandings of "Nature,"[3] many of which are supposed to be much more eco-friendly than the exploitative attitudes underlying modern capitalist, consumerist societies. In many remaining pockets of "traditional societies," there are divine trees, sacred groves, and reverential attitudes toward trees. Sometimes these reverential attitudes and practices toward trees are being held up as possible models for emulation, the necessary warnings against naïve romanticism notwithstanding. China, too, has been seen as possessing some traditional eco-friendly attitudes toward "Nature" and associated cultural practices. The Daoist philosophy of "no action" (*wuwei* 無為) is understood to at least potentially lead to cultural practices that have less impact on the environment (see Girardot et al. 2001; Miller 2017; also Miller et al. 2014). Despite the Daoist anthropocosmic resonance and Buddhist compassion-toward-all-living-things models of nature, it was the Confucian technocratic and utilitarian model that prevailed in traditional Chinese attitudes toward, and actions upon, the environment. This model was further strengthened by the import of post-Enlightenment European conceptions of humanity-over-nature, culminating in the environmentally disastrous high-socialist and high-modernist state-led industrialization and development schemes (see Shapiro 2001; Weller 2010).

Robert Weller's work (2010) reminds us that whatever eco-friendly conceptions and practices there might be in traditional Chinese religious traditions, these are not in themselves "environmentalist" in the sense this word means today. He suggests that while we parochialize Euro-American conceptions of nature and environmentalism (i.e., challenging their hegemony), we need to take care not to fall into the trap of exoticizing some imagined Oriental wisdoms of nature. Weller has crucially distinguished "environmental cultures" from "environmentalism." By "environmental culture," Weller refers to ideas and practices relating to the environment that are not necessarily, or even often, environmentalist (just like "Nature," what constitute "the environment" depend on cultural and sociopolitical contexts; see Brown's chapter in this volume). In the Chinese context, such environmental culture could include, as discovered by Weller, displaying a "scholar's rock" on one's writing desk (complete with tiny bonsai trees as miniature cypress trees and figurines in "mountain grottoes"); mixing nature tourism with traditional forms of religious pilgrimage or fun-centered family outings (what Weller calls "mélange tourism"); picking up strangely shaped or patterned pebbles from nature reserves and placing them on ancestral altars; deities speaking through spirit mediums against industrial developments; and fights between communities as well as against the state over where a garbage incinerator should be located.

What happens when different environmentalisms encounter one another, or when a particular local "environmental culture" encounters more hegemonic forms of environmentalism (see Horowitz et al. 2017)? More specifically, what happens when different regimes of tree-mindedness encounter one another? Things can get quite unpredictable and interesting when the already-diverse environmental ideologies and cultures in different societies (Euro-American, Chinese, Japanese, or any other) encounter one another in a thoroughly globalized and marketized world (Weller 2010). The "friction" created by these encounters in various "zones of awkward engagement" (Tsing 2005) sparks multi-stranded and multi-layered effects. "Negotiation" has been a key word in the anthropological literature in characterizing these encounters. I am suggesting two more metaphors: "intermeshing" and "mutual capture." Intermeshing refers to the coming together of two or more relatively well-formed "regimes" (of tree-mindedness, of production, of revolutions, etc.), as if two or more meshes (like sieves) are brought to bear upon one another. Some stuff "stick" while others pass through the holes. The stuff that stick are the sociopolitical and cultural materials that get elaborated further (though one may never know if some of the stuff that don't stick this time might come back and stick later). "Mutual capture" refers to the ways in which various social agents and institutions snatch up elements from local as well as translocal and even global pools of resources (symbolic or material) for their own agendas, all the while themselves getting captured in return. I will elaborate on these two metaphors further in the conclusion.

When tree-mindedness has become such a pervasive phenomenon in China (or for that matter in so many other parts of the world), it became inevitable that an otherwise non-tree-minded anthropologist such as myself ran into many tree-minded people in my field (e.g., Shaanbei) and the interesting things these people do with trees. I was "in-*tree*-gued" (apologies for this silly pun!).

In the introductory section, I have briefly mentioned the "global-environmentalist regime of tree-mindedness," which should be familiar to most readers. The rest of this chapter has three substantive sections and a conclusion. The following, second, section is a short presentation of some of the features of tree-mindedness in traditional "indigenous" Chinese cultural contexts, for example, *fengshui* considerations, the role of the magistrate in encouraging tree-planting; trees in and around temple grounds. The third section presents a few features of the Chinese "socialist-mobilizational regime of tree-mindedness." This includes *danwei* (單位 work unit)-based tree-planting; the Three-Planting Day and tree-planting as a civic duty; tree-planting by the People's Liberation Army; tree-planting and corporate responsibilities; the development of forestry science; and the large-scale reforestation of agricultural land.[4] The fourth section shows how different regimes of tree-mindedness intermesh and interact using the concrete ethnographic case of the reforestation project at the Black Dragon King Temple in Shaanbei, northcentral China.[5] I will first recount how the concern for legitimacy motivated the temple to take up the idea of building a Hilly-Land Arboretum from a local forestry expert. Then I describe and analyse a tree-planting activity initiated by Friends of Nature (*ziranzhiyou* 自然之友), a well-known Chinese environmentalist NGO based in Beijing, and hosted by the temple. The final section offers a reflection on the connections among local cults, environmentalism, and Chinese cosmic-governance. In the conclusion, I will elaborate on "intermeshing" and "mutual capture" and draw out some of the implications of the findings.

II. TREE-MINDEDNESS IN TRADITIONAL CHINESE CULTURAL CONTEXTS

Before I introduce the much more modern regimes of tree-mindedness, I will first briefly present the various cultural conceptions and practices relating to trees in China.

Given the vastness of China's expanse and long history, tree-related symbolic constructs are numerous in number and wide ranging in referent domains (e.g. society, politics, religion, aesthetics, philosophy, health and vitality, education, personal character, and construction). One particular feature of these constructs is the frequent use of the word tree (*shu* 樹) as a verb to mean "to erect," "to establish," and "to form" (e.g., to "cultivate talents" *shuren* 樹人). It would be fair to say that the Chinese have always been tree-minded in one way or another.

In southern China, where *fengshui* (風水) discourse is much more elaborate than in the north, especially in relation to topographical features. There are communal or lineage *fengshui* forests that are quite similar to what's called "sacred groves" in studies on South Asia (Gold and Gujar 1989; 2002). In Shaanbei (northern Shaanxi Province), I have not come across similar concepts, even though *fengshui* considerations by hired *fengshui* masters (called *yinyang xiansheng* 陰陽先生 or *pingshi* 平士 locally) are routine for the construction of dwellings for both the living and the dead.

In dynastic times, peasants mostly considered trees as economic resources (e.g., as sources of firewood; timber for construction of houses and temples; and producers of fruits, tea leaves, or leaves for silkworms and domestic animals). Some mountain regions with navigable waterways became prized commercial logging farms whose access rights were jealously guarded by powerful local lineages (see McDermott 2013 and 2020 on the Huangshan area in late-imperial Huizhou). The magistrates or prominent members of the local gentry had to persuade or order people to plant trees in public spaces (e.g., along roads, dykes). People would not normally initiate these public works projects themselves since these could not be harvested by individual households (figure 8.1).

Temples and monasteries in traditional times often had extensive land holdings, whose agricultural produce supplied the resident clerical communities and whose income supported temple activities as well as the maintenance and expansion of temple buildings and grounds. Some of the temple lands were wooded (especially if the temples were in or near some hills), and the

Figure 8.1 **A panoramic view of a big valley near the County Seat of Hengshan in Shaanbei.** The tall trees are found along main roads, rivers, and irrigation canals, and they normally belong to the village collective or the highway bureau maintaining the main roads. Photo by author, 2016.

trees supplied fruits, medicine, firewood and timber, not to mention a habitat for game (e.g. deer, rabbits, wild boars, and birds) and even mystical retreats.

III. SOCIALIST-MOBILIZATIONAL REGIME OF TREE-MINDEDNESS

The Maoist state organized large-scale tree planting as part of its "fight against nature" (Shapiro 2001), for example, against soil erosion and desertification, and in order to "greenify the mother land" (*lühua zuguo* 綠化祖國). Commune members were organized by their production teams to plant trees along the fields and in the hills when they were not busy with agricultural work or political studies, and urban work units were given tree-planting tasks in the countryside around the city (more on this below).[6]

But what modern environmentalists understand to be "the environment" ([*ziran*] *huanjing* [自然]環境) and the ideal environment socialist tree-planting is supposed to help achieve are two entirely different things (and concomitant realms of sociopolitical practices): one is about the protection of supposed pristine nature at risk of human encroachment while the other is about the ever intensification of human engagement and transformation of deficient or threatening "nature" (it doesn't matter whether the source of this deficiency or threat was natural processes or anthropogenic) (figure 8.2).

Danwei-Based Tree Planting

The most interesting phenomenon relating to socialist tree-mindedness was the organization of urban work units, or *danwei*, to plant trees every year. This form of tree planting was specific to the high-socialist era (1950s to 1970s). Every spring, hundreds of thousands of *danwei* workers (schools, academic research units, factories, hospitals, banks, etc.) would be put on Liberation-brand, military-green, open-top trucks and brought to the peri-urban country-side to plant trees. These were well-coordinated military-style exercises. The tree saplings and required equipment (wheel-barrows, spades, digging hoes, buckets, etc.) would be ready for them at the designated sites and there would be precise and scientifically rationalized division of labor and work flows. These were mostly day trips as the *danwei* workers arrived at the site in the early morning and returned home in the city at night. They might continue like this for a few days and then be taken over by another group of work units.

Figure 8.2 Chinese propaganda poster from the Maoist era rallying the whole nation to plant trees. Tree-planting was labeled as a "campaign." *Source*: The IISH / Stefan R. Landsberger / Private Collection (https://chineseposters.net/posters/pc-197b-001)

The work was hard, especially for urban dwellers not used to physical labor. But these were also seen as pleasurable occasions for mass sociality (there was much singing, slogan-shouting and high-spirited comradery).

During the high-socialist era, all activities were part of the centralized, redistributive political economy, and all resources, be they land, labor, raw materials, finished products, were "collective" (except personal belongings), which allowed the different work units to coordinate large-scale collective activities such as construction and agricultural work, transportation, and, important to our story in this chapter, tree planting. The accumulation of cognitive and organizational know-how around planting trees and ensuring their survival on a massive scale during the high-socialist period formed the basis of reform-era tree-planting mobilizations.

The Tree-Planting Day and the Making of Environmental Subjects

During the reform period in the past four decades, the high-socialist redistributive political economy devolved into a mixed political economy, following privatization of a large portion of the state sector and the development of the market. Work units no longer organize any tree-planting or any other similar campaign-style activities (though most still organize group tours for their employees and retirees during official long holidays). Instead, all citizens are encouraged to plant trees as a civic duty (*gongmin de yiwu*). The Tree-Planting Day (*zhishujie*) was instituted in 1979 to encourage all citizens to plant trees.[7]

The Tree-Planting Day connects nation-building, the political socialization of the citizenry by the modern nation-state, and tree-planting activities (see Agrawal 2005 on "environmental subjects"). However, it is almost impossible for urbanites in today's China to find a suitable place to plant trees (all empty plots of land in or near the cities are either owned by developers or belong to the municipal government under some kind of urban planning). It has become logistically challenging for an average urbanite family to locate a suitable place to plant trees (often involving commercial tour companies). This is an important factor in understanding the desire and necessity for urban-based environmentalist NGOs to make special arrangements with social organizations in the countryside (e.g., temples) in order for their members to be able to plant trees.[8]

Tree-planting, the Army, and Corporate Responsibilities

Meanwhile, the Chinese state is stepping up its tree-planting activities, but in a different guise. For example, platoons of People's Liberation Army engage in tree-planting activities on a regular basis (just as they are deployed in disaster relief efforts). In recent years, tens of thousands of soldiers have been deployed to plant trees to fight desertification and air pollution in northern China. A recent campaign in 2018 involved assigning 60,000 soldiers to plant trees covering an area in northern China the size of Ireland with an aim to combat pollution and sand storms.[9]

The Chinese state also encourages corporations (including those state-owned enterprises that have evolved from their *danwei* days) to engage in planting trees as part of their corporate responsibilities program (CRP). As a result, many corporations build long-term collaborations with certain rural townships, especially those in "poverty counties," in order to plant trees or sponsor the planting of trees with monetary input. It is small wonder that the most enthusiastic corporate participants in this scheme are companies involved in infrastructural work (construction companies), given their existing technical strength and proximity to large infrastructural projects that demand accompanying environmental "beautification" (e.g., green belts along highways and around airports and high-speed-train stations).[10] On the other hand, ordinary non-corporate institutions, such as universities, secondary and primary schools, hospitals, research institutes, and so on, do not have the human and financial resources to arrange long-term tree-planting commitments.

The Forestry Bureau and the Science behind Tree-Planting

One crucial development of the Maoist period was the establishment and expansion of the Forestry Bureau (*linyeju* 林業局) (now called the State Forestry and Grassland Administration of China).[11] The *danwei*-organized tree-planting activities were most often coordinated with local forestry bureaus that provided the tree saplings from their nurseries. The forestry bureaus were instrumental in scientific research on what species of trees were most suitable for which kinds of climate, soil, and other conditions, and what kinds of planting techniques were crucial for the best chance of survival for the trees. In the story on the establishment of the Hilly-Land Arboretum at the Dragon King Valley, we will see the key role played by a particularly enterprising local forestry researcher (a certain Mr. Zhu) and the legitimation role played by the national forestry experts and institutions.

Reverting Farmland to Forests

In the past two decades or so, the Chinese government initiated a policy of reverting farmland to forests (*tuigeng huanlin* 退耕還林) in order to encourage peasants to reduce their farming land coverage so that some of the land, especially those higher-altitude and ecologically fragile slope lands, can be reforested (e.g., through abolishing the agricultural tax and giving villagers subsidies). This was a nation-wide policy, and all of a sudden there was a frenzy to replant trees in fallowed farmlands (mostly on hills and slopes). As a result, the business of tree saplings was all the rage since there was limited supply of tree saplings by state forestry nurseries. In Shaanbei, many peasants began growing tree saplings as a cash crop in their best, irrigated fields in the flatlands in the hope of cashing in onto this market. However, as is often the case, when too many people had the same idea, the outcome was the rapid reduction in sale price, to a point when the sale price was so low that it was not even worth selling them. The tree saplings continued to grow (very fast! talk about affordance!) and soon they became too big and no one wanted to expend the considerable energy and money (since one needs to hire some machinery) to chop the trees down, dig up the roots, and dispose of the whole thing. The unintended consequence of this miscalculation and agricultural minor disaster is the "greenification" of vast expanses of farmland. Perhaps these trees will eventually be cut down and sold as timber but one needs to wait and see. This mixture of elements of state policy and commercial adventurism with its unintended consequences constitutes an

Figure 8.3   Tree farms on irrigated land (Shengshan County, Shaanbei). Photo by author, 2016.

interesting episode in the socialist-mobilizational regime of tree-mindedness (figure 8.3).

IV. TEMPLE-BASED REGIME OF TREE-MINDEDNESS AND THE POLITICS OF LEGITIMATION

The Black Dragon King Temple (Heilongdawangmiao 黑龍大王廟) in the Dragon King Valley (Longwanggou 龍王溝) in Shaanbei (i.e., "my" temple! see Chau 2006) began a reforestation project in the mid-1980s. Now hundreds of acres of hilly land around the temple are covered with a thick blanket of trees (mostly evergreens) (officially called the Longwanggou Hilly-Land Arboretum, *shandi shumuyuan* 山地樹木園), making it one of the most green and beautiful sites in Shaanbei (see Chau 2018).

But why is there anything particularly striking about planting trees on or near a temple in China? Isn't it true that traditionally, many Buddhist monasteries, Daoist shrines, as well as popular religious temples had magnificent trees in their courtyards, around the parameters of the temple ground, or had a special grove of trees in the back out of *fengshui* considerations? Isn't it the case that some temples were even specifically devoted to the worship of tree spirits, which granted protection to some individual trees or groves of trees? Does not Daoism in particular advocate the harmony of people and nature/cosmos? It is indeed very tempting to interpret the Longwanggou reforestation and botanical project as an expression of a sort of Chinese folk environmentalism, as a continuation of a long tradition of the close connection of Chinese religious cosmology to nature and cosmic forces. However, a closer look at the history and the operations of the arboretum will reveal that the picture is far more complicated.

Using materials collected in the mid-1990s in rural northcentral China (Shaanbei, northern Shaanxi Province), supplemented with more recent documentary materials, I shall in this section of the chapter show how the Black Dragon King Temple succeeded in tapping into the expanding environmentalist discourse in China and grafting a tree-planting enterprise (the Hilly-Land Arboretum) onto the existing temple complex, thus attracting visitors to the temple site from far and wide, including even environmentalist-activists and students from Beijing (eighteen hours away by chartered bus). This is a story of the transformation of a site that changes the overall configuration of the sitescape by incorporating elements that are new, which, in turn, attract new categories of site visitors (i.e., tree-planters or *arbortourists*) (see Chau 2018). Indeed, many different kinds of social actors were involved in bringing together the disparate domains that are religion, forestry, environmentalism, and schools.[12]

The Politics of Legitimation: The
Longwanggou Hilly Land Arboretum

By the 1990s (my fieldwork was between 1995 and 1998), Longwanggou
(literally the Dragon King Valley, which stands for the Black Dragon King
Temple) had firmly established itself as one of the most successful temples
in Shaanbei in terms of fame and donation income (second only to the
much more prominent White Cloud Mountain, Baiyunshan 白雲山 Daoist
Complex in Jia County, Jiaxian 佳縣).[13] The temple's success was primarily
thanks to the charismatic leadership of its temple boss, Mr. Wang Kehua (or
Lao Wang, "Old Wang," an affectionate way to refer to a middle-aged man
in China), a villager from one of the three core villages running the temple.[14]
Each year in the sixth month of the lunar calendar, the annual temple festival
attracted a couple of hundred thousand visitors over the course of a few days,
making it one of the most red-hot event productions in the Shaanbei region.[15]
The temple became very rich and began engaging in a variety of activities
such as running a primary school affiliated to the temple (later on even a
secondary school), sponsoring irrigation projects and engaging in charity
work. In the mid-1980s, Mr. Zhu Xubi, a native of Zhuzhai, one of the six
new affiliated villages to Longwanggou, came to temple boss Lao Wang with
an idea. Zhu Xubi was a forestry engineer working at the Yulin Prefecture
Forestry Scientific Research Institute (*linkesuo* 林科所, a shortened form of
linye kexue yanjiusuo 林業科學研究所) and he proposed to Lao Wang to use
some of the temple donation money to start a reforestation and botanical project at Longwanggou. At first Lao Wang was not very interested and treated
Zhu Xubi as one of the many people who pestered him for money. But Zhu
eventually persuaded Lao Wang to experiment with the idea.

A meeting was held with those Hongliutan villagers who have land around
the temple. The temple offered to lease their land to plant trees in return for
a rent that is the equivalent of the estimated highest yield possible from those
plots.[16] Most of these plots are on steep or rocky slopes, have no irrigation
access, and are thus agriculturally unproductive. The most common crops for
this kind of marginal farm land are potatoes and black beans with low yield.
Hence, it was not difficult to persuade the Hongliutan villagers to lease out
these plots to the temple for a steady annual rent. At this time, many Hongliutan villagers were involved in trade, commerce, or rural industry and were
only happy to have their farming obligations reduced. In addition, not farming the land also meant that they would no longer need to pay agricultural
tax on those plots (this tax was later abolished altogether). Another meeting
was held among all the temple officers and village representatives to the
Longwanggou temple association to approve the use of temple funds on the
arboretum project.

Meanwhile, the project had to be approved by the relevant state authorities. Like many local state service type agencies (*shiye danwei* 事業單位), the Yulin Prefecture Forestry Scientific Research Institute was extremely short of funds. As a result, many of its personnel became idle (because no work was properly remunerated) and not much was going on. Zhu was one of the few more enterprising employees and hit upon the idea of using temple funds to plant trees. In June 1988, Zhu obtained permission from the institute's supervisory agency, the Yulin Prefecture Forestry Bureau (*linyejü*), to officially make the Longwanggou Arboretum into a research project (*lieru keti* 列入課題) to include many possible forestry-related scientific experiments. In November, the same year, the Yulin Prefecture Forestry Bureau, the Zhenchuan Township government, and Longwanggou Historical Relic Management Office cohosted a workshop (*yantaohui* 研討會) consisting of sixty-seven experts and scholars from twenty-seven work units in forestry, scientific research, historical relic management (because at that time the Black Dragon King Temple was officially registered as a historical relic management unit), and culture to discuss the issues related to the establishment of the arboretum. Longwanggou of course footed the bill to host, accommodate and fête lavishly all the guests. At the conclusion of the meeting, the Longwanggou Civil Hilly Land Arboretum (*minban shandi shumuyuan* 民辦山地樹木園) was officially established ("civil" because it is not operated by the state). Since the arboretum project would not be possible without the money from incense donations to the temple, the arboretum (just as the primary school affiliated with the temple) therefore justified and shielded the divine efficacy-related activities of the Heilongdawang Temple such as divination and the dispensing of "magical healing water" (officially considered as superstitious and illegal) from possible official criticism and crack down.[17] It was a manifestation of the politics of legitimation (commonly seen in all similar situations of negotiating acceptance under unfavorable legal-political environments in semi-authoritarian regimes).

After having secured the land for the project and the green light from the authorities, the next step was to plant trees. Capitalizing on his long years of experience as a forestry engineer and his personal and professional connections to different forestry bureaus, seedling stations and botanical gardens in Shaanbei and beyond, Zhu Xubi managed to obtain for Longwanggou a large number of different species of trees, shrubs, and flowers. The temple truck became very handy in hauling the plants back to Longwanggou. The planting process was an arduous task, as the land was not suitable for trees and needed a lot of preparation. Lao Wang and the other temple officers worked with a team of hired laborers and volunteers to blast holes into the rocky slopes, build small terraces, haul dirt and seedlings up the hills and plant the trees (see figure 8.4). Zhu and his associates provided the scientific

Figure 8.4   The Black Dragon King Temple Festival in the summer of 2016. The main temple hall is on the left. The temple dormitory and temple-run primary school are in the building on the right. Many new features have been added since the 1990s (e.g., the pavilion on one of the hilltops). The lush green scenery is very much evident in the surrounding hills, all part of the hilly-land arboretum. Photo by the author, 2016.

expertise in zoning the different species and ensuring the optimal conditions for survival for the new trees. The arboretum subsequently also served as a scientific research base for the Yulin Forestry Scientific Research Institute with dozens of different research projects. A few of these projects won some prefectural level scientific research awards. Zhu Xubi was allocated an office at the temple and became a resident consultant for the arboretum.

The significance of the arboretum project has to be seen in the larger context of rising national environmental concerns in the reform era and Shaanbei's harsh environmental setting. Yulin county and many of Shaanbei's northern counties face constant threats of desertification and dust storms (part of the Maowusu Desert of Inner Mongolia extends southward into Shaanbei), and because of the long history of human settlement and dry climate, Shaanbei has a severe tree shortage. As a result, reforestation has always been a high-profile issue in Shaanbei. The Longwanggou Arboretum project capitalized on the moral virtue of environmentalism—this was around the time when environmentalism became all the rage in public discourse on a national level in China—and thus quickly gained regional, national and even international attention. One cannot underestimate the aesthetic and emotional appeal of the green hills around the temple, set against the often parched landscape of Shaanbei loess plateau. In 1996, Longwanggou hosted another workshop where even more officials, scientists, and scholars converged to assess the arboretum's achievements. The main guests at this workshop were representatives of the official China Botanical Institute (*zhongguo zhiwu xuehui* 中國植物學會), who confirmed the arboretum's environmental and scientific value through the ritualistic activity of official endorsement by certificate granting.

Planting Trees as "Doing Religion"

In examining the activities engaged in by Longwanggou, it is useful to evoke the concept of "doing religion" (see Chau 2006). Even though the Heilong-dawang Temple is theoretically a "religious" institution, its actual activities have gone far beyond the religious. Therefore, it is not adequate to only consider the state's religious policies and examine how local temples react to these policies. Instead, it is important to look at what local temples and local state agencies do. The Heilongdawang Temple's various religious and non-religious domains of activities constitute how local people "do religion" on the ground. In these activities, the actions of local state agents and sometimes social actors from far away (such as the tree-planting visitors I will discuss below) are as important as the villagers and temple organizers.

The temple's many-stranded activities have induced various branches of the local state to descend upon Longwanggou. Of course, sometimes Long-wanggou invited the presence of these local state agencies out of concerns of legality and legitimacy. As these vertical ties between different local state agencies and Longwanggou multiply and thicken, the legitimacy of the temple increases, as no single local state agency alone can determine the fate of the temple (not even the Religious Affairs Bureau, which became relevant after the temple was officially granted the status of a "venue for religious activi-ties" in 1998). This functional expansiveness of popular religious temples demonstrates that the emerging "religious field" in today's China is often shot through with institutional arrangements and practices that are nonreligious, and often these nonreligious aspects play the role of legitimating the religious aspects of the temples. This is particularly true in the case of popular religion as compared to the officially recognized five religions because of the former's status as theoretically illegal feudal superstitious activities.

The local state's regulatory paternalism is, on the other hand, a form of local state activism, when local state agents also actively seek ways to expand their resource and profit base. For example, the establishment of the Longwanggou Hilly Land Arboretum exemplifies the negotiation, mutual

cooptation and cooperation between Longwanggou and those local state
agencies in the forestry sector (even if the forestry expert Mr. Zhu was ini-
tially not necessarily representing the interest of the local state).

The idea of *minjian* (民間 nongovernmental) has gained salience in
public discourses in the PRC since the 1980s. While the idea of "people"
(*renmin* 人民) still retains a lot of currency (as an object to be acted upon
by the state or to be served by "civil servants"), the idea of *minjian* points
to an expanding public sphere where citizens act upon their own initiatives
(see Yang 1994). Lao Wang and his associates learned to promote the high
appeal of the fact that the Longwanggou Hilly Land Arboretum was the
first minjian hilly land arboretum in China—this "firstness" is a significant
symbolic capital in today's China, signifying an innovative spirit. In fact,
the twin features of *minjian*-ness and "folk environmentalism" exemplify
two of the most important ideological emphases of the reform era: privatization and environmental protection. This has lent Longwanggou an aura
of legitimacy and savvy of which other temples are rightfully jealous.
Newspapers reported on the arboretum; officials and foreign dignitaries
visited it; botanists, forestry specialists and other scientists came to bestow
their approval and marvel at this folk initiative; environmentalist groups
from Beijing and NGO groups from Japan came to plant trees (more on
this below). The media was particularly happy to have discovered a great
story: peasant intellectual (Lao Wang) turning superstitious activities into
a tree-planting project that benefits the people. In a fashion reminiscent
of "model-making" (*shuli dianxing* 树立典型) during the Maoist era, Lao
Wang was made into a folk hero whose vision, dedication and leadership
transformed the badly eroded barren slopes of Shaanbei into a green and
beautiful oasis.

The presence of foreigners in particular has bestowed added legitimacy
upon Longwanggou. In the summer of 1990, Longwanggou received the first
foreign visitors who had heard about the temple and the arboretum. A few
Japanese were the first to come, and they were followed by a steady trickle of
foreigners of different nationalities. Not many foreigners had visited Shaanbei because much of it was still officially closed to foreigners at that time,
therefore their visit to Longwanggou conferred importance and recognition
on Lao Wang and his colleagues' enterprise. Quite a few of these foreigners
were journalists, scientists, or academics and they wrote about Longwanggou
after they went back to their countries. Through one of these transnational
connections, Lao Wang was invited to become a Chinese member of Inter-
Asia, a Japanese-funded, pan-Asian, NGO. As the leader of Longwanggou,
Lao Wang was invited to participate in national and even international conferences on forestry, environmentalism, and sustainable development. The visits
of foreigners to Longwanggou became so frequent that the Yulin Prefecture

Foreign Affairs Office (*waishiban* 外事辦) decided to grant Longwanggou official permission to host and register foreign visitors.

The success of the arboretum was a huge boost to the official status and image of the temple, and the temple has been riding on this success ever since. In the 1990s, a few other Shaanbei temples had followed suit to use temple funds to initiate reforestation projects. The most famous one among these is in Shilisha (十里沙 Ten Mile Sand), also in Yulin County, based on a temple dedicated to the Perfected Warrior Ancestral Master (*Zhenwuzushi* 真武祖師). The arboretum at Shilisha was subsequently consecrated as China's first civil *sandy* land arboretum (*minban shadi shumuyuan* 民辦沙地樹木園) after going through a process of accreditation similar to the one I just described for Longwanggou. In the past two decades or so, his arboretum trend has spread in Shaanbei and other places as it has provided popular religious temples a legitimation model to emulate.

An interesting twist in the establishment of the Shilisha Civil Sandy Land Arboretum is that it was again Zhu Xubi the forestry engineer who initiated it. Upset at what seemed to be Zhu's lack of loyalty to Longwanggou, Lao Wang eventually banished Zhu from Longwanggou and replaced him with a younger forestry technician. This demonstrated how conscious Lao Wang was about the potential competition for fame and recognition other temple-cum-arboretums might present. Zhu was very bitter about the experience, but sought comfort in his other, new arboretum sites. He was also bitter that now Lao Wang received all the credit for conceiving and starting the arboretum project and no one knew or cared about his crucial contributions.

Lao Wang's temple boss status and his vision for the temple's development and the uses of temple funds have not gone uncontested. Village temples are in principle public goods that belong to the village collectivity. Yet when there is village factionalism, members of opposing factions would vie for control of the temple association the way they would the Villagers' Committee. Temples serve as yet another battleground for local political maneuvers and intrigues. Because temple offices are public, the cultural ideal for a temple officer is caring for the common good without partiality. In reality, however, all recognize the inevitability of temple officers' serving their particularistic interests. Given the internal divisions of every community, a temple officer is bound to be perceived as a benefactor and ally by some people and a bully and enemy by others.

With or without the temple association, micro-politics is endemic in village communities. A loose coalition of Lao Wang's enemies had formed over the years and they launched one attack after another in their attempt to dislodge him from power. Among other attacks, they sued him for taking valuable farm land to build the new school building for the Longwanggou Primary School (located on flat irrigable land that used to be the Hongliutan village orchard) and the Longwanggou Arboretum. This attack was particularly damning, as the then (late 1990s) newly propagated national Land Law (*tudifa* 土地法) stipulated strict approval requirements for converting farm land to nonagricultural uses. One man who felt particularly wronged by Lao Wang accused him publicly in the form of Maoist-era denunciation posters. Here are just a few lines excerpted from the rather lengthy handwritten denunciation text:

> Wang Kehua sits in the Dragon Valley. He is famous in the region and has his gang in the county. *He uses the forest park to protect the god, the god to protect himself and the money to protect his power.* He has been in control of the temple for eighteen years. To enrich himself he has done so many things. He has a corrupted morality. He has done rotten and criminal deeds to oppress the masses and to spoil the party and the government. . . . *You do things that violate the country's laws. Incense temple and forest park? What will humans depend on for food? You use money to buy fame! That forest park of yours? There are too many people and too little land in Hongliutan. How much flat land and hilly land per head? Turning huge pieces of good farm land into forestry land for the god?* (heavily redacted and italics added).

Among other things, this man rather perceptively pointed out the intricate role the arboretum plays in protecting the temple and the role temple funds play in protecting Lao Wang. Lao Wang had to face counter-legitimation challenges constantly, and his enemies often resorted to law (in this case the Land Law) to delegitimize his project and his temple-boss status. Because of Lao Wang's extensive connections with different local state agencies and officials at the township, county and prefectural levels, he was able to successfully weather most of these challenges. But he did admit to me once that he was afraid that one day some "bad elements" might set fire to the arboretum and destroy the fruits of so much effort.

During my fieldwork period at the Dragon King Valley, I did not hear the local villagers discuss the arboretum in religious terms. They came to appreciate the beauty of the reforested hilltops and slopes around the temple, but no one said explicitly that the trees performed any role such as improving the *fengshui* of the site. However, there is one concrete link between the trees and funerals. In Shaanbei as in many other parts of China, people would build a "spirit shed" (*lingpeng* 靈棚) to house the coffin during the funeral (see Chau 2006 chapter on hosting funerals). Due to the traditional symbolic connections between longevity and evergreen trees, especially cypress (*bai* 柏), people often decorate the spirit shed with stems of cypress leaves. The attendants of the Hilly-Land Arboretum told me that they sometimes got requests from villagers in the surrounding areas for cypress branches to decorate spirit sheds when there were deaths in their households. They could not say no to these requests, so they would cut off branches from different cypress trees in the arboretum and gave them to these villagers. Since a large proportion of the trees in the Arboretum were various kinds of drought-resistant cypresses, presumably meeting these occasional requests did not pose any threat to the health of the trees. I do not know if Longwanggou still gets these requests (presumably yes). This service to the local community is certainly an unexpected outcome of the tree planting efforts at Longwanggou (perhaps we can call this the "funerary regime of tree-mindedness"?).

Longwanggou Hosts a Tree Planting Event:
The Visit of Beijing Arbortourists

One event related to the arboretum that I witnessed and participated in during my fieldwork in 1998 is a particularly good example to illustrate how the temple and Lao Wang tried to capture "the powerful outside" to bolster the temple's legitimacy, how metropolitan and global environmentalism articulated with folk environmentalism, and how a certain kind of tourism (what I call arbortourism) interfaced with a religious site despite the fact that the site-visitors ostensibly ignored the religious attributes of the site all the while with the full corroboration of their local hosts who also elected to only highlight the trees and downplay the temple. As a result of these efforts of mutual dissimulation, a sort of theater was staged, with a happy outcome for both parties.

The leaders of Friends of Nature, a Chinese environmentalist organization NGO based in Beijing, had heard about the achievements of the Longwanggou Arboretum project and decided to bring some of their members as well as some Beijing secondary school students to Longwanggou to plant trees (tree-planting being one of their most important outreach activities). The director of the organization was a certain Dr. Liang Congjie, a university professor and a well-known member of the National Political Consultative Conference (*zhengxie* 政協). As Lao Wang was elected to be a member of the Yulin County Political Consultative Conference, Dr. Liang and he were also linking up as members of the same nation-wide organizational system (*xitong* 系統), though at two very different levels (one national while the other merely local). Lao Wang the local elite had control over some local resources (i.e., Longwanggou) to offer the national elite (Dr. Liang and his associates) in exchange for possible favors in the future. It was trees that the Beijingers were coming to Longwanggou to plant, whereas Lao Wang and his associates wished to "plant" crucial *guanxi* (connections) with some potentially useful external forces. The Beijingers were to come in the spring of 1998, during the May Day long vacation (*wuyi laodongjie* 五一勞動節).[18]

Lao Wang was very excited about their imminent visit. He made various arrangements to ensure that their visit would be pleasant and successful. Of course, no one can just show up at a place and decide to plant some trees; a lot of preparatory work is needed. At that time, all of the hilly land around the temple had been planted, so Lao Wang ordered the forestry technician and a team of laborers to vacate a couple of patches of the arboretum not too far from the temple to let the Beijingers plant trees. This meant digging up many healthy trees that had already taken root to some other spots in the arboretum at the risk of killing them. They also needed to purchase tree saplings from some nearby sapling stations and store them at the temple until the day the Beijingers came. Other things they had to purchase were shovels, buckets, hoses, and tags for the Beijingers to mark the trees they would plant as "their trees" so that in the future they could come back and check up on them. The forestry technician warned that it would not be the perfect time to plant those trees (more than a month later than the ideal planting time) and he had serious doubts that they would survive. But the "tree-planting event" as a performance had to succeed, even if the new saplings might not survive and some old trees had to be dug up to make room for some cosmopolitan environmentalist arbortourists.[19]

The members of Friends of Nature and the Beijing students and teachers came in three big rented sleeper buses in the early morning of May 1st, after about eighteen hours of travel. They even brought a TV crew and a few Chinese and foreign journalists. There were altogether about 120 visitors. They were welcomed warmly by the temple staff and fed a hearty and multicourse Shaanbei breakfast (presented as "revolutionary food" though considerably superior to the real revolutionary diet of the Red Army or that of the high Maoist era). After resting a little while in their dormitory rooms that had been specially readied for them, they climbed up the hill behind the temple with the tools and began planting trees under the supervision and with the help of Longwanggou Arboretum and temple staff. It seemed to be a fun activity for the Beijingers, especially the students; they were singing songs, chatting and laughing. But all worked hard. For them this was the most suitable activity for a "labor day."

The tree-planting activity occupied the entire first day. On the second day, the Beijingers visited the nearby village Batawan on some kind of folk culture tour; the students were assigned to different peasant households to experience Shaanbei life. This kind of rural-life tourism was only beginning to take shape in China in the 1990s. It soon took on the umbrella term "peasant family happiness" (*nongjiale* 農家樂) (see Park 2008; Chio 2012). Other than some minor incidents all the activities went smoothly.[20] In the evening, a "happy-together" gala (*lianhuan wanhui* 聯歡晚會) was held on the opera stage where the Beijing visitors and the Longwanggou Primary School students performed songs, dances, *yangge* (秧歌 a group dance style in local folk tradition), and skits to a large and enthusiastic audience that was composed of villagers from the area.[21] The gala was concluded by a large *yangge* dance in the open area in front of the opera stage where all the Beijingers participated.

The next morning, before the Beijingers were to depart in their buses, a little solemn ritual was held. A beautiful stone stele was officially "revealed" (*jiebei* 揭碑) at a prime spot near the temple with inscriptions to commemorate the tree-planting event (the polished black granite stone alone cost the temple 1,000 yuan) (see chapter 11 in Chau 2006 for the "juicy" story behind the erection of this stele). The stele was also intended to provide material expression of Longwanggou's far-flung translocal ties and legitimacy. Curiously, the traditional idiom of erecting a stele at a temple ties Longwanggou's arboretum enterprise and the Beijingers' arbortourism back to popular religion (all temples feature steles commemorating the rebuilding or renovation of various temple structures; see Chau forthcoming).

The success of this tree-planting event provided Longwanggou yet another organizational idiom or model to interact with outsiders (the other idiom being the experts' workshop I mentioned earlier). In terms of monetary and labor input, the tree-planting event was very taxing for Longwanggou, but it apparently was deemed worth the effort, at least from Lao Wang's perspective. Thereafter the Friends of Nature came to Longwanggou for more tree-planting activities. Members of Inter-Asia, the Japanese-funded, pan-Asian, NGO to which Lao Wang belongs also came to Longwanggou to plant trees. More commemorative steles were erected.

What is most interesting about these arbortourists' visits to Longwanggou is *how seemingly far tree-planting as an environmentalist-driven activity is from religion.* The Beijinger arbortourists seemed to have fastidiously avoided touching anything religious when they were at Longwanggou (most likely because an understanding was reached among the adults, including the teachers, that it would be better to play safe and not to expose the schoolchildren in their charge to "feudal superstition"). It seems that as far as they were concerned, they were visiting the Longwanggou Hilly Land Arboretum and not the Black Dragon King Temple, and the trees they were planting were to benefit the environment conceived broadly rather than the Black Dragon King Temple grounds. This *conceptual apartheid between the environment and religion* was in fact crucial to the use of the arboretum and the arbortourists in legitimating Longwanggou, which until 1998 was not officially registered as a venue for religious activities, which meant that all the religious activities taking place at the temple (dispensing divine healing water, divination, worshiping and making offerings to the Black Dragon King, providing child-protection rituals, the temple festival, etc.) were illegal "feudal superstition" that should have been suppressed by the Public Security Bureau (i.e., the police). But all the money (lots of money!) that sustained these tree-oriented activities (including the hiring of forestry technicians and laborers; irrigation works; the buying and transportation of saplings and tree-planting equipment; the hosting of arbortourists, forestry experts, journalists and other site-visitors; the rental on the plots; etc.) came from incense donations to the temple, which ultimately were derived from folk religiosity and the belief in the Black Dragon King's magical efficacy, his miraculous response to the prayers of the worshippers (even if no direct link was claimed between the Dragon King and the trees except the latter's role as provider of water). If the Beijingers and other visiting arbortourists were genuinely ignorant of the religious nature of the funding that enabled their visits and their tree-planting activities, then they seemed to have come to this encounter with Longwanggou in bad faith (though I am certain some were not so naïve). Lao Wang and the temple, on the other hand, knew perfectly well what they were doing.

V. LOCAL CULTS, ENVIRONMENTALISM, AND CHINESE COSMIC-GOVERNANCE

Some readers might be disappointed that there are no concrete instances of religiously inspired environmentalism in my account. The Black Dragon King Temple did not have a tradition of planting trees around the temple. For hundreds of years, the impetus for the local villagers was always to maximize agricultural output for subsistence and, if there was any surplus, for the market (not so much for profit as for cash to pay taxes). As a result of this agrocentrism, all arable land was used for farming.[22] In addition, most of the slopes around the temple were simply too rocky to sustain trees naturally. For the tens of thousands trees of the arboretum to survive, the irrigation water does not come from the Black Dragon King's spring, which is only a small trickle; nor does it come from the sky due to the chronic drought condition all of northern China faces; instead, the temple initiated substantial waterworks to ensure a constant supply of water (including hooking up to, and paying for, the public water supply provided by the County Water Bureau). For the temple, the reforestation project was first and foremost a legitimation strategy, while the happy outcome of beautifully reforested slopes and hilltops around the temple is mostly an unintended consequence of this legitimation politics. Once established, the reforestation project took on a life of its own, bringing all kinds of dividends to the temple that the original initiators could not possibly foresee. In other words, one could say that the more precarious the status of an institution (temple cult as "feudal superstition"), the more likely that it engages in activities that will help it gain ready approval and legitimacy in the eyes of wider society and the authorities (e.g., opening a school, planting trees, engaging in charity, etc.). This same logic seemed to be operating in the establishment of the "Daoist Ecological Forest" (*daojiao shengtailin* 道教生態林) in Minqin County, Gansu Province, where a relatively weak Daoist community latched onto the "Daoist Ecological Forest" project that was a political assignment from the county government in the 1990s in order to raise the status of Daoism locally (Yang Der-ruey's chapter 7 in this volume).

Maintaining the hilly-land arboretum is expensive (rents to the villagers whose plots have been taken away, paying for water from the public water supply, hiring laborers to plant trees and water them all year round, buying tree saplings, paying consultation fees to forestry experts, hosting tree-planting activities, etc.). Only very successful temples such as the Black Dragon King Temple have the funds (from its substantial incense-money donations that amount to millions of Chinese dollars each year) to support such an operation. The lack of a constant source of adequate funds was the primary factor in the eventual failure of the Daoist Ecological Forest project in Gansu (Yang Der-ruey this volume). Perhaps this financial aspect is one of the secret (yet obvious) links between religion and environmentalism in contemporary China: temples bring in money (which is based on people's belief in the divine efficacy of the deities) and part of this money supports environmentalist projects. But this link is not based on either religious teachings or prior practices; rather, it has resulted from the intermeshing of different regimes of tree-mindedness and the mutual capture of cognitive, material, institutional, and other resources (more on this in the Conclusion).

Another secret (yet also somewhat obvious) link between religion and environmentalism as exemplified in the Longwanggou case is the primacy of the locale. The Black Dragon King cult is a local cult, revived and managed by local villagers. For forty years, the control over the temple and its funds has stayed firmly in the hands of the locals. The success of the temple brings fame and material benefits to the villagers (despite some intermittent struggles between different factions) as well as the whole region. Even if the locals do not see the reforested slopes and hilltops in religious terms, many of them have become appreciative of the aesthetic appeal of the overall improved landscape (a sensibility, I must add, that was not necessarily one of a traditional Shaanbei peasant). This landscape, despite its all-season evergreenness, is resolutely local, as the reforested areas covered and followed familiar topographical features of the area. The "work" of the arboretum has contributed to the construction of Longwanggou's overall "ritual *terroir*," that is, a kind of locale-based vitality (see Chau 2020). If there is any religiously inspired ethics that is relevant for environmentalism, it would be this commitment to the locale that lays at the foundation of all local cults (even if this commitment itself is not "religious" in nature).

Due to the limits imposed by these concerns for the locale (e.g., Temple Boss Lao Wang getting jealous that another temple would become a competitor in reforestation efforts and potentially stealing the limelight), the kinds of environmentalist projects such as the Longwanggou Arboretum seem to be difficult to scale up, except as being replicated in other similar locales in equally local dimensions. This seems to be the limitations of the potential impact of local cults on the environment. However, the Chinese state can be seen as similarly concerned with the locale, only this "locale" is the Chinese nation. Tristan Brown's work (this volume and under review) shows clearly that the late-imperial Chinese state was very much into entertaining local people's appeals against encroachments onto their "physical" surroundings (e.g., mining) based on *fengshui* principles because the late-imperial state itself was constituted in parallel terms. Brown has called this state an "ecological polity" or "eco-polity," highlighting the importance of "the environment" to the health, vitality, and survival of the polity (especially the ruling house). Building upon Brown's insight, I recognize this "environment" as conceived in cosmic terms, including not only the natural and built environments, but even more importantly the invisible forces of virtue, *fengshui*, divine blessings, demonic spirits, karmic recompense, and cosmic resonance.

The modern Chinese state has in important ways inherited this model of what can be called "cosmic-governance," much modified vocabulary notwithstanding. The reduction of the "All under Heaven" (*tianxia* 天下) scope to the boundaries of the modern nation-state in fact aided the sacralization of China as the "sacred/divine realm" (*shenzhou* 神州), as a "ritual *terroir*" (e.g., *shenzhou dadi* 神州大地) to be further worked on. The mystified fervor and ritualized practices in "conquering" and domesticating "Nature" under Chinese socialism as well as the mass-mobilizational call to "greenify China's sacred/divine realm" (*lühua shenzhou dadi* 綠化神州大地) are parallel to the imperial state's attempt at cosmic-governance, as both are grounded in deep-seated cosmological considerations (especially regarding the role of the state in effecting the best possible welfare for the [Chinese] people). We must repeatedly remind ourselves that for the Chinese state and its ruling elites, "the environment" is primarily about Chinese people's "living environment" (*shengcun huanjing / shenghuo huanjing* 生存環境 / 生活環境) rather than the environment in contemporary environmentalist discourse. Even though in recent years the Chinese government is widening its environmentalist concerns to a global scope (e.g., signing international environmentalist agreements combating climate change), its proposal of a so-called ecological civilization (*shengtai wenming* 生態文明) (see Hansen et al. 2018; Coggins 2019) or "green civilization" (*lüse wenming* 綠色文明) is principally about improving *China*'s environment for the benefit of *Chinese* people. Environmentalism has become an important instrument in Chinese statecraft and official nationalism, and tree-planting (or "greening") its linchpin and somewhat fetishized practice. However, is this locale-centrism or even the "étatization by ecology" a bad thing (Litfin 1997 speaks of the "greening of sovereignty")? Or can the state work with, and "resonate" with, much smaller-scaled terroirs such as Longwanggou to achieve better environmentalist goals?[23]

VI. CONCLUSIONS: INTERMESHING OF
REGIMES AND MUTUAL CAPTURING

I would like to suggest the notions of "intermeshing" and "mutual capturing" to best understand what have been going on at Longwanggou with temple bosses, local villages, forestry experts, tree-planting practices, arbortourists, and so on. As a particular regime of tree-mindedness develop over time, it is inevitable that it comes into contact with other regimes of tree-mindedness, and always in concrete settings, whether it is in an Amazonian or Kalimantan forest (see Tsing 2005) or at a temple not far from the advancing edges of the Ordos Desert. These encounters often result in competition, contestation, mutual borrowing, and contingent intermeshing. For example, the global-environmentalist regime of tree-mindedness came to China starting in the 1990s and helped produce urban-based environmentalist NGOs (e.g., Friends of Nature) that necessarily inherited practices from Chinese socialist-mobi-lizational regime of tree-mindedness (most of the founders of these NGOS were former *danwei* employees, and they often had to strategically play by rules set by the Party-state). As these different regimes of tree-mindedness intermesh, sometimes a few conceptions or forms of activities stand out as being acceptable to all parties. These are the elements that "stick" and become further elaborated into more enduring practices. Tree-planting is one of these elements nobody can object to (but see the Longwanggou case given earlier for instances of objections from some locals who believe in the virtue of a tilled and agriculturally productive land, rather than some reforested woodland serving no apparent purpose other than legitimizing a temple cult and protecting the temple boss).

Besides "intermeshing," I am also proposing the metaphor of "mutual capture." It was Deleuze and Guattari (1987) who first introduced the concept of "apparatuses of capture," but the ritual studies scholar Kenneth Dean reworked the concept in the context of analyzing the complex ritual networks and "ritual machines" in southeastern coastal China (Dean 1998: 45; also Dean and Zheng 2010). In explicating the exuberance of ritual performances on the Fujian coast (across the Taiwan Strait from Taiwan), Dean defines "apparatuses of capture" as "the capture and temporary consolidation of social, economic, political, and libidinal forces by cultural forms" (Dean 1998: 45). In the aforementioned extended ethnographic vignette, we see plenty of examples of how the folk (i.e. the villagers sponsoring the Black Dragon King Temple) captures useful elements of the state (i.e., local state agencies, scientific expertise, and endorsement) and transnational environmentalism; and how agents of the state captures the folk to further their ends; and how metropolitan and transnational environmentalism captures the folk.

In the environmental anthropology literature, we often read about examples of how religion is used to protect forests in the form of *fengshui* forests or sacred groves or wooded areas preserved for the purpose of providing forest products for religious events or safeguarding good *fengshui* (e.g., Gold and Gujar 1989; Tsu 1997; Coggins 2014; 2019). The Longwanggou case suggests yet another way in which religion and trees interact: in this case, trees are used to protect the temple politically and in turn the temple's income and religio-physical location (i.e., the temple *as a site*, see Chau 2018)] sustain the trees and the improved "ritual terroir."[24] The Longwanggou case study should, I hope, illustrate even more clearly how the different kinds of environmentalisms, or more specifically and accurately, tree-mindedness, converge and articulate with one another. The tree-mindedness of the members of Friends of Nature and the Beijing students has been nurtured by global and state environmentalist discourse, NGO practice, tree-planting practices, and so on. Had the temple not captured the expanding environmentalist discourse, taken advantage of (socialist) scientific forestry expertise, and founded the Hilly-Land Arboretum, it would never have been able to capture the elite site-visitors such as the Friends of Nature environmentalist-activists and the students and teachers of an elite secondary school in Beijing. I believe "mutual capture" is a useful notion in analyzing these situations, as these sites and "site-visitors" (Chau 2018) are normally very far apart from each other but somehow each has managed to capture the other for one's own agendas.

NOTES

1. Evans-Pritchard famously referred to the Nuer as "cattle-minded," which caused him to become cattle-minded as well in order to understand them (1940: 242).

2. See Mueggler (2011) on Western plant hunters in southwestern China in the early twentieth century and Hemery (2019) on a Victorian plant hunter's adventures in North America in the mid-1800s. There was certainly a regime of tree-mindedness in the imperialist era of competitive botanical science when every newly discovered and named botanical species (trees included of course) was glory for the empire or colonial power sponsoring the expedition.

3. I put "nature" in quotation marks because it is a sociohistorical construct.

4. There are other regimes of tree-mindedness in China (e.g., late-imperial; funerary; corporate-commercial), but we will not be considering these in this article.

5. With the revival of religious life in China beginning in the early reform era (early 1980s), many temples that had been destroyed in the high socialist era have been rebuilt, especially in the countryside. The Black Dragon King Temple is one of the most well-known temples in Shaanbei and was the focus of my doctoral research (see Chau 2006). This case study on the NGO tree-planting activity has been featured in various forms in a number of my past publications (Chau 2006; 2009; 2018). In this article I will use the case study to support a very different argument.

6. During the Maoist era (and much less so today), work units (*danwei*) were socialist "total institutions" that provided not only employment, housing, food and other welfare facilities (e.g., nurseries, clinics, and sports fields), but also political indoctrination and mobilization.

7. It was the Republican government that first instituted the tree-planting day to commemorate the founder of the Republic, Dr. Sun Yat-sen.

8. More recently urbanites are encouraged to "adopt" existing trees in their neighborhoods (i.e., donating an annual amount to help maintain the greenery in the surrounding areas).

9. https://www.independent.co.uk/news/world/asia/china-tree-plant-soldiers-reassign-climate-change-global-warming-deforestation-a8208836.html.

10. Here are some examples of corporate tree-planting activities: https://www.bcegc.com/menu374/newsDetail/1893.html; http://www.sdsdjt.com/?list_4/321.html.

11. See Sivaramakrishnan (1999) for a study on the role modern forestry played in the construction of modern polities.

12. I have taken most of the ethnographic description from my monograph *Miraculous Response: Doing Popular Religion in Contemporary China* (Chau 2006) and have expanded on some sections. Most of the ethnographic materials were collected during eighteen months of fieldwork in Shaanbei (northern Shaanxi Province in north-central China) between 1995 and 1998. Administratively, Shaanbei comprises Yan'an and Yulin prefectures. My principal fieldsite, the Dragon King Valley, is located in Yulin County, which is one of the twelve counties belonging to the Yulin Prefecture. The loess plateau and cave dwellings of Shaanbei were made famous when the Central Red Army, led by Mao Zedong, made Yan'an the capital of their revolutionary base area from 1935 until 1945 (my fieldsite in Yulin was under Nationalist rule during that period). Shaanbei has been one of the poorest regions of China until coal and natural gas were found in the northern parts of Shaanbei, whose exploitation is bringing more prosperity to the region.

13. Popular religion has enjoyed a momentous revival in China during the reform era (1980's onward), especially in rural areas. New temples have been built and old temples restored; local opera troupes crisscross the countryside performing traditionally themed opera pieces for deities and worshippers at temple festivals (*miaohui*); *fengshui* masters (geomancers) are busy siting graves and houses and calculating auspicious dates for weddings and funerals; spirit mediums, Daoist priests, gods, and goddesses are bombarded with requests to treat illnesses, exorcize evil spirits, guarantee business success, and retrieve lost motorcycles; before anti-Falungong suppressions began in 1998, *qigong* (氣功) sects of all manners competed for followers not only in the cities but in rural areas as well.

14. The temple committee has representatives from nine villages surrounding the temple, but three of these villages have dominant positions due to their traditional claim to the temple. Mr. Wang Kehua was instrumental in the revival and rebuilding of the temple in the early 1980s (when popular religion revived in many parts of China) so was chosen as the temple boss (more by general inclination than formal election). The temple boss heads the temple committee and has autocratic power in decision making.

15. Red-hotness (*honghuo/re'nao*) is the ideal condition for occasions such as temple festivals and wedding/funeral banquets. Event productions are socially staged activities (e.g., temple festivals) that are more complex and less coherent than rituals. For more on red-hot sociality and event productions, see chapters 7 and 8 in Chau (2006) and Chau (2012).

16. All land in China officially belongs to the state. During the reform period (post early 1980s), villages are given the right to distribute plots to households depending on the number of household members, but people only have use right rather than ownership over these plots. The use right of the land surrounding temples in the Chinese countryside is the key to any local tree-planting efforts. Unlike in the pre-PRC days, temples today do not own any land, and their tree-planting or other activities that will take up any village land will depend on the villagers' good will. No urban *danwei* or enterprises have use right to this kind of land due to the strict urban-countryside division instituted during the Maoist period, and when urban institutions impinge upon rural areas it's only for developmentalist goals (factories and apartment blocks) rather than "useless activities" such as planting trees. None of the revived urban temples (very few in number) have land outside of their immediate grounds to plant trees.

17. As the local state has, during the reform era, become accommodationist and regulatory vis-à-vis popular religion, there have been very few incidents of official crackdown on temple cults. But the pursuit of official recognition was essential for warding off possible future episodes of official persecution.

18. Intending to stimulate consumption and leisure activities, the reform-era Chinese state expanded the May First International Labor Day into a long vacation that is enjoyed by school children and employees of state and private enterprises alike. The other long vacation is during the Chinese New Year. Both can last up to ten days. To go on a tree-planting expedition during vacation is very much a legacy of the Maoist era, when different urban work units and schools organized collective volunteer labor activities such as planting trees and rice seedlings, harvesting, and building dams and terraced fields.

19. As if blessed by the Heilongdawang, most of the trees actually survived through at least the summer of 1998, thanks to the abundant spring rain, the availability of irrigation and the care of the arboretum staff.

20. A female Beijing student was bitten by a peasant home guard dog (she was trying to take a picture with the dog thinking that it was a pet while in fact peasants use dogs primarily as guard dogs and they can be extremely fierce). She was rushed to the local hospital to be checked if she had contracted rabies (fortunately she did not). One of the Friends of Nature members was known to be physically weak but she worked very hard until she got sick and had to lie down in the dormitory.

21. Some of the Longwanggou Primary School children's parents complained that the temple and Lao Wang were exploiting their children in entertaining the Beijing guests. But the Beijingers did reciprocate appropriately by donating a large number of books to the schools of the nine villages affiliated with Longwanggou. The majority of the books went to the Longwanggou Primary School. See Chau (2007) for an analysis on youth and youth cultural production in rural China.

22. When a traditional peasant looked at a piece of land, the first thing that came to his mind would be if this could be turned into a plot, if it would be irrigable, which crops could be grown here, and so on.

23. This question is quite the opposite in spirit to those proposals that wish to downplay the role of the nation and nationalism (e.g., Duara 2014).

24. A recent example of religion tapping into environmentalist discourse is the Green Pilgrimage Network started in 2011 that includes four Daoist sites in China: http://greenpilgrimage.net/members/china/louguantai/. Accessed May 4, 2016. I thank Knut Aukland for this information.

ENGLISH AND WESTERN LANGUAGES

Agrawal, Arun. 2005. *Environmentality: Technologies of Government and the Making of Subjects*. Durham, NC: Duke University Press.

Brosius, J. Peter. 2003. "The Forest and the Nation: Negotiating Citizenship in Sarawak, East Malaysia." In *Cultural Citizenship in Island Southeast Asia: Nation and Belonging in the Hinterlands*, edited by Renato Rosaldo, pp. 76–133. Berkeley: University of California Press.

Brown, Tristan. Under Review. *The Law of the Living Earth: Fengshui, Property, and Environment in Late Imperial China*. Book manuscript.

Chau, Adam Yuet. 2006. *Miraculous Response: Doing Popular Religion in Contemporary China*. Stanford: Stanford University Press.

Chau, Adam Yuet. 2007. "Drinking Games, Karaoke Songs, and Yangge Dances: Youth Cultural Production in Rural China." *Ethnology*, vol. 45, no. 2, p. 161–172.

Chau, Adam Yuet. 2009. "Expanding the Space of Popular Religion: Local Temple Activism and the Politics of Legitimation in Contemporary Rural China." In *Making Religion, Making the State: The Politics of Religion in Contemporary China*, edited by Yoshiko Ashiwa and David Wank, pp. 211–240. Stanford University Press.

Chau, Adam Yuet. 2012. "Actants Amassing (AA)." In *Sociality: New Directions*, edited by Nick Long and Henrietta Moore, pp. 133–155. Berghahn Books.

Chau, Adam Yuet. 2018. "Of Temples and Trees: The Black Dragon King and the Arbortourists." *International Journal of Religious Tourism and Pilgrimage*, vol. 6, no. 1, Article 8, pp. 72–84.

Chau, Adam Yuet. 2019. "Hosting as a Cultural Form" *L'Homme* (Special issue "Cumulus: Hoarding, Hosting, Hospitality"), no. 231–232, pp. 41–66.

Chau, Adam Yuet. 2020. "Ritual Terroir: The Generation of Site-Specific Vitality." *Archives des Sciences Sociales des Religions* (Special Issue on Réguler

les pluralités religieuses : mondes indiens et chinois comparés / regulating religious pluralities: comparing the Indian and Chinese worlds. Edited by Vincent
Goossaert).

Chau, Adam Yuet. Forthcoming. "Temple Inscriptions as Text Acts." In *Text, Context, and Acts: Chinese Popular Religion in Practice*, edited by Shin-yi Chao.
Amsterdam: University of Amsterdam Press.

Chio, Jenny. (Director) 2012. *Peasant Family Happiness* (Nong Jia Le). Ethnographic film. 71 minutes.

Coggins, Chris. 2014. "'When the Land is Excellent': Village Fengshui Forests and
the Nature of Lineage, Polity, and Vitality in Southern China." In *Religious Diversity and Ecological Sustainability in the People's Republic of China*, edited by J.
Miller, D. Yu and P. van der Veer. New York: Routledge.

Coggins, Chris. 2019. "Sacred Watersheds and the Fate of the Village Body Politic in
Tibetan and Han Communities Under China's *Ecological Civilization*." *Religions*,
vol. 10, no. 11, p. 600.

Dean, Kenneth. 1998. *Lord of the Three in One: The Spread of a Cult in Southeast
China*. Princeton, NJ: Princeton University Press.

Dean, Kenneth and Zheng Zhenman. 2010. *Ritual Alliances of the Putian Plain. Volume One: Historical Introduction to the Return of the Gods*. Leiden: Brill.

Deleuze, Gilles and Félix. Guattari. 1987 [1980]. *A Thousand Plateaus: Capitalism
and Schizophrenia*. Translated by B. Massumi. Minneapolis: University of Minnesota Press.

Duara, Prasenjit. 2014. *The Crisis of Global Modernity: Asian Traditions and a Sustainable Future*. Cambridge: Cambridge University Press.

Escobar, Arturo. 1995. *Encountering Development: The Making and Unmaking of the
Third World*. Princeton: Princeton University Press.

Girardot, Norman J., James Miller and Liu Xiaogan, eds. 2001. *Daoism and Ecology:
Ways Within a Cosmic Landscape*. Cambridge, MA: Harvard University Press.

Gold, Ann Grodzins and Bhoju Ram Gujar. 1989. "Of Gods, Trees and Boundaries:
Divine Conservation in Rajasthan." *Asian Folklore Studies*, vol. 48, no, 2, pp.
211–229.

Gold, Ann Grodzins and Bhoju Ram Gujar. 2002. *In the Time of Trees and Sorrows:
Nature, Power, and Memory in Rajastan*. Durham, NC: Duke University Press.

Greenough, Paul and Anna Lowenhaupt Tsing, Eds. 2003. *Nature in the Global
South: Environmental Projects in South and Southeast Asia*. Durham, NC: Duke
University Press.

Hansen, Mette Halskov, Hongtao Li and Rune Svarveruda. 2018. "Ecological Civilization: Interpreting the Chinese Past, Projecting the Global Future." *Global Environmental Change*, vol. 53, pp. 195–203.

Hemery, Gabriel. 2019. *Green Gold: The Epic True Story of Victorian Plant Hunter
John Jeffrey*. London: Unbound.

Horowitz, Leah S. and Michael J. Watts, eds. 2017. *Grassroots Environmental Governance: Community Engagements with Industry*. London: Routledge.

Litfin, Karen T. 1997. "Sovereignty in World Ecopolitics." *Mershon International
Studies Review*, vol. 41, no. 2, p. 167–204.

McDermott, Joseph P. 2013. *The Making of a New Rural Order in South China Volume 1: Village, Land, and Lineage in Huizhou, 900–1600.* Cambridge: Cambridge University Press.

McDermott, Joseph P. 2020. *The Making of a New Rural Order in South China Volume 2: Merchants, Markets, and Lineages, 1500–1700.* Cambridge: Cambridge University Press.

Miller, James. 2017. *China's Green Religion: Daoism and the Quest for a Sustainable Future.* New York: Columbia University Press.

Miller, James, Dan Smyer Yu and Peter van der Veer, eds. 2014. *Religion and Ecological Sustainability in China.* London: Routledge.

Mueggler, Erik. 2011. *The Paper Road Archive and Experience in the Botanical Exploration of West China and Tibet.* Berkeley, CA: University of California Press.

Naquin, Susan and Chün-fang Yü, Eds. 1992. *Pilgrims and Sacred Sites in China.* Berkeley: University of California Press.

Park, Choong-hwan. 2008. "Delights in Farm Guesthouses: *Nongjiale* Tourism, Rural Development and the Regime of Leisure-Pleasure in Post-Mao China." PhD dissertation University of California, Santa Barbara.

Rival, Laura, ed. 1998. *The Social Life of Trees: Anthropological Perspectives on Tree Symbolism.* New York: Berg.

Seeland, Klaus, ed. 1997. *Nature is Culture: Indigenous Knowledge and Sociocultural Aspects of Trees and Forests in Non-European Cultures.* London: ITDG Publishing.

Shapiro, Judith. 2001. *Mao's War against Nature: Politics and the Environment in Revolutionary China.* Cambridge: Cambridge University Press.

Sivaramakrishnan, K. 1999. *Modern Forests: Statemaking and Environmental Change in Colonial Eastern India.* Stanford: Stanford University Press.

Slater, Candace, ed. 2003. *In Search of the Rain Forest.* Durham, NC: Duke University Press.

Tsing, Anna Lowenhaupt. 2005. *Friction: An Ethnography of Global Connection.* Princeton, NJ: Princeton University Press.

Tsu, Timothy. 1997. "Geomancy and the Environment in Premodern Taiwan." *Asian Folklore Studies*, vol. 56, no. 1, pp. 65–77.

Weller, Robert P. 2010. *Discovering Nature: Globalization and Environmental Culture in China and Taiwan.* Cambridge: Cambridge University Press.

Index

human ethics-ism (*rendaozhuyi*,
人道主義), 92
human exceptionalism, 45
humanism (*rendao zhuyi*), 88–89, 92,
192. *See also* human ethics-ism
humanism: exclusive, 92, 96, 123;
Renaissance, 89
human rationality, 35
human rights, 35–37, 44–46, 60;
Christian anthropocentric provenance
of, 20, 47; Enlightenment notion
of, 9, 21, 59; inalienability of, 39;
institution of, 56–57; in Taiwan's
constitution, 61; in Taiwan's Life
Education curriculum, 57. *See also*
natural rights
Hunan, 104, 112
hun-po (魂魄, cloudsouls and
whitesouls), 49–50, 52–53, 153–54.
See also spirit (*shen*)
husheng (protection of life), 21, 41,
43–44, 46, 176
hydraulic engineering, 7, 13

inanimate synthetic objects, 21, 27,
84–85
indigenous: life ways, 19; ontologies,
15, 58; sensibilities, 19;
understandings of Nature, 226;
wisdom, 18, 142
indigenous people, 15, 19, 51–52;
consumption of, 51–52; Taiwan, 51
industrial: development, 1, 18, 123,
125–26, 128, 227; production, 7, 80
industrialization, 6, 72, 191, 226–27
industrialized, 7, 126–27
Industrial Revolution, 6–7, 84
institutionalized activism, 141
intermeshing, 228–29, 247, 249
International Confucian Ecological
Alliance (ICEA), 140
International Humane Society (IHS),
189, 191
International Network of Engaged
Buddhists (INEB), 36

Jamaica Bay Wildlife Refuge, 170, 174,
192

Kangxi, 110
karma: collective (*gongye*, 共業),
48, 55, 81–83, 86, 96; collective
(*gongye*, 共業) *vs.* individual (*duye*,
獨業), 81–83; positive and negative,
183; transmigration and, 54. *See also*
Orbits of Releasing Life
karmic response, 175
Kohn, Eduardo, 13, 15, 58, 152, 154,
155, 158–59, 161

Lambek, Michael, 9, 11
Laozi (老子), 49, 123
li (禮, ceremony), 162
Life Conservationist Association (LCA)
(關懷生命協會), 35, 52
linguistic anthropology, 11
living environment, 189, 248
locale, 200
locale-based vitality, 247–48
locale-centricism, 26
Locke, John, 20, 35, 39–41, 45, 56,
60–61
Lockean human equality, 36, 42, 44
Lockean rights, 37
longmai (龍脈). *See* dragon vein
Longwanggou (Dragon King Valley),
25, 233
Longwanggou Hilly-Land Arboretum,
25, 235–53

Mahā prajnāpāramitā śāstra
(《大智度論》), 176
mai (脈). *See* veins
Maoist, 5, 26, 91, 226, 230; era, 4, 7–8,
90, 91, 231, 240, 242, 244, 252
Mark, Robert, 23
Master Jingyi (釋淨義), 172–73
May Day, 243
Meaning of Sacrifice (*jiyi*), 49
mechanistic Newtonian universe, 14
meridian (*mai*), 20, 107. *See also* veins

Contributors' Biographies

Tristan G. BROWN 張仲思

Tristan G. Brown (PhD in history, Columbia University) is assistant professor in the Department of History at MIT. His research interests include the history of the Ming and Qing Dynasties, environmental history, legal history, as well as the history of Chinese borderlands. He has been a junior research fellow in Asian and Middle Eastern studies at St. John's College, Cambridge University, and a Chinese studies postdoctoral fellow at Stanford University. He has published widely on topics of Chinese law, environment, and religion.

Robert Ford CAMPANY 康若柏

Robert Ford Campany (PhD in history of religions, University of Chicago) is professor of Asian studies and religious studies at Vanderbilt University. He works on the history of religions in late classical and medieval China as well as on the history of the cross-cultural study of religion. To date, he has authored six books: *Strange Writing: Anomaly Accounts in Early Medieval China* (1996); *To Live as Long as Heaven and Earth: A Translation and Study of Ge Hong's Traditions of Divine Transcendents* (2002); *Making Transcendents: Ascetics and Social Memory in Early Medieval China* (2009); *Signs from the Unseen Realm: Buddhist Miracle Tales from Early Medieval China* (2012); *A Garden of Marvels: Tales of Wonder from Early Medieval China* (2015); and *The Chinese Dreamscape, 300 BCE–800 CE* (2020). He has also coedited two volumes, and published over forty articles and chapters.

Adam Yuet CHAU 周越

Adam Yuet Chau (PhD in anthropology, 2001, Stanford University) is reader in the anthropology of China, teaching in the Department of East Asian Studies, University of Cambridge, and is a fellow at St. John's College. He is the author of *Miraculous Response: Doing Popular Religion in Contemporary China* (Stanford University Press, 2006) and *Religion in China: Ties That Bind* (Polity Press 2019), and the editor of *Religion in Contemporary China: Revitalization and Innovation* (Routledge, 2011). His edited volume *Chinese Religious Culture in 100 Objects* is yet to be published. He is interested in developing better ways of conceptualizing Chinese religious culture. One of his out-reach ambitions is to stop people from asking the question "How many religions are there in China?" His other book projects investigate the idiom of hosting (*zuozhu*) and forms of powerful writing ("text acts") in Chinese political and religious culture.

HUANG Weishan 黃維珊

Huang Weishan (PhD in sociology, New School for Social Research) was an assistant professor in the Department of Cultural and Religious Studies at Chinese University of Hong Kong. Her work mainly focuses on new religious movements, religion and migration, civil society, and religion and gentrification. She is coeditor of the book, *Ecology of Faith in the New York City* (Indiana University Press, 2013). Her current research is on the reconfiguration of two significant state-planned social phenomena, urbanization and religious revival, and its impacts on Mahayana Buddhist communities in contemporary Shanghai. She is currently Associate Professor in the Sociology Department at Hong Kong Shue Yan University.

LIANG Yongjia 梁永佳

Liang Yongjia is professor and director of the Institute of Anthropology, Zhejiang University, China. He is interested in Sinophone anthropology and the ethnographic study of religion and ethnicity in China and the Asia-Pacific. He publishes on secularism, religious revival, and ethnic identity. His journal articles were published in *Asia Pacific Journal of Anthropology*; *China: An International Journal*; *Religions*; *China Review*; and *CARGO*. His monograph *Religious and Ethnic Revival in a Chinese Minority* was published by Routledge in 2018. Presently, he is working on interpreting non-Chinese societies with Chinese concepts.

Jeffrey NICOLAISEN 倪杰

Jeffrey Nicolaisen (PhD in religion, 2019, Duke University) is currently a Duke-DKU Global Fellow at Duke Kunshan University in Kunshan, China, and has recently served as both a research associate and visiting instructor at Duke University and Duke Kunshan University. His research uses Taiwanese traditions and teachings to rethink networks of human and nonhuman agency and the ethics of multi-species interaction between Han and indigenous people, dogs, and monkeys in Taiwan. Nicolaisen was a

Fulbright-Hays Fellow at Taipei Medical University in 2017–2018 and a Charlotte Newcombe Fellow in 2018–2019 at Duke University. Prior to pursuing an academic career, he worked as an environmental consultant with the global sustainability consulting group Environmental Resources Management.

WEI Dedong 魏德東

Wei Dedong (PhD in Chinese philosophy, Renmin University of China in Beijing) is professor of the School of Philosophy, researcher of the Institute for the Study of Buddhism and Religious Theory, and the Director of the International Center for Buddhist Studies at Renmin University of China. His main research areas are in Chinese Buddhism and the Sociology of Religion. He is the founding editor of *Series of Buddhist Translations, International Buddhist Studies* and *Translations in the Classics of the Sociology of Religion*. He served as a primary investigator of the China Religion Survey at Renmin University from 2011 to 2020. His publications include *The Essence of Buddhist Yogacara Philosophy*, and three volumes of the *Dedong Review on Religion*.

YANG Der-Ruey 楊德睿

Yang Der-Ruey (PhD in anthropology, London School of Economics and Politics) is a professor in the Institute of Social/Cultural Anthropology, Nanjing University, China. His research mainly concerns the transmission of religions, especially Daoism and the folk religiosities of the Han Chinese people. He is also interested in the transmission of traditional Chinese arts and crafts, such as herbal medicine, folk music, and calligraphy. His publications include *Transmission: An Anthropological Exploration of Cognition and Religion* (The Commercial Press, 2018), "Revolution of Temporality: The Modern Schooling for Daoist Priests in Contemporary Shanghai" in Liu, Xun & David Palmer, eds. *Between Eternity and Modernity: Daoism and its Reinventions in the 20th Century* (UC Press, 2012), "From Crafts to Discursive Knowledge: How Modern Schooling Changes the Learning/Knowledge Style of Daoist Priests in Contemporary China" in Chau, Adam, ed. *Religion in Contemporary China: Revitalization and Innovation* (Routledge, 2010).

Mayfair Mei-hui YANG 楊美惠

Mayfair Mei-hui Yang (PhD in anthropology, UC Berkeley) is professor of religious studies and East Asian studies at the University of California, Santa Barbara. Her areas of research include: modernity, religiosity, and secularization; critical theory; China studies, sovereignty, and state power; gender and feminism; media studies; and political economy. She has published two monographs: *Gifts, Favors, and Banquets: the Art of Social Relationships in China* (1994, Cornell University Press, American Ethnological Society Prize); and *Re-enchanting Modernity in China: Ritual Economy and Religious Civil Society in Wenzhou* (2020, Duke University Press), and edited two volumes: *Chinese Religiosities: Afflictions of Modernity and State Formation* (2008, University of California Press); and *Spaces of Their Own: Women's Public Sphere in Transnational China* (1999, University of Minnesota Press). She also made the documentaries: *Through Chinese Women's Eyes* (1997, distributed by Women Make Movies), and *Public and Private Spheres in Rural Wenzhou, China* (1994). She has published many articles in journals such as *Current Anthropology, Journal of Asian Studies, Annales, Comparative Studies in Society & History, China Quarterly,* and *Public Culture.* He was also the director of Asian studies at the University of Sydney in Australia (2007–2009). Her new monograph *Religious Environmentalism in the Age of the Anthropocene: Potentialitites and Actualities in China* is yet to be published.

www.ingramcontent.com/pod-product-compliance
Lightning Source LLC
Chambersburg PA
CBHW050409280326
41932CB00013BA/1798